GENDER AT TH.

BORDER REGIONS SERIES

Series Editor: Doris Wastl-Walter

In recent years, borders have taken on an immense significance. Throughout the world they have shifted, been constructed and dismantled, and become physical barriers between socio-political ideologies. They may separate societies with very different cultures, histories, national identities or economic power, or divide people of the same ethnic or cultural identity.

As manifestations of some of the world's key political, economic, societal and cultural issues, borders and border regions have received much academic attention over the past decade. This valuable series publishes high quality research monographs and edited comparative volumes that deal with all aspects of border regions, both empirically and theoretically. It will appeal to scholars interested in border regions and geopolitical issues across the whole range of social sciences.

Gender at the Border
Entrepreneurship in Rural Post-Socialist Hungary

JANET HENSHALL MOMSEN
University of California, Davis, USA

with

IRÉN KUKORELLI SZÖRÉNYI
*Centre for Regional Studies of the Hungarian Academy of Sciences,
Győr, Hungary*

and

JUDIT TIMAR
*Centre for Regional Studies of the Hungarian Academy of Sciences,
Békéscsaba, Hungary*

Routledge
Taylor & Francis Group

LONDON AND NEW YORK

First published 2005 by Ashgate Publishing

Reissued 2018 by Routledge
2 Park Square, Milton Park, Abingdon, Oxon OX14 4RN
711 Third Avenue, New York, NY 10017, USA

Routledge is an imprint of the Taylor & Francis Group, an informa business

First issued in paperback 2018

Typeset by J.L. & G.A. Wheatley Design, Aldershot, Hampshire

A Library of Congress record exists under LC control number: 2005929791

Notice:
Product or corporate names may be trademarks or registered trademarks, and are used only for identification and explanation without intent to infringe.

Publisher's Note
The publisher has gone to great lengths to ensure the quality of this reprint but points out that some imperfections in the original copies may be apparent.

Disclaimer
The publisher has made every effort to trace copyright holders and welcomes correspondence from those they have been unable to contact.

ISBN 13: 978-0-815-38916-3 (hbk)
ISBN 13: 978-1-138-61973-9 (pbk)
ISBN 13: 978-1-351-15768-1 (ebk)

Contents

List of Figures

List of Tables

List of Plates

Acknowledgements

We were persuaded to undertake this collaboration by Dr György Enyedi, Vice-President of the Hungarian Academy of Sciences, and are very grateful for his support throughout the project.

Irén Kukorelli's husband, Miklós Szörényi, is a statistician, and was an enormous help in analysing the questionnaires. He was also very patient with our comings and goings and supportive of the time we have spent on this work. We thank him for all his help.

Maps and figures were drawn by Margit Szőke Bencsikné, cartographer at the Centre for Regional Studies of the Hungarian Academy of Sciences in Békéscsaba. We are grateful for the speed at which she put our ideas on to paper and how efficiently she dealt with working in a language she did not understand.

We also thank Doris Wastl-Walter, series editor, for her support for our research and her encouragement for its presentation in this book. It was Doris who first got us interested in the geopolitical nature of borders, especially in a changing Europe. Finally we thank Val Rose of Ashgate who has been very patient and understanding throughout the genesis of the book and has also provided several practical suggestions along the way.

We dedicate this book to György Enyedi in recognition of all the work he has done for Hungarian geography and for his encouragement to the three of us over many years.

Chapter 1

Introduction

As the borders of the European Union changed with the addition of ten new members on 1 May 2004, an understanding of the contrasting economies of the pre- and post-expansion borderlands became of wider interest. Of these ten countries two are Mediterranean islands (Malta and Cyprus), three are Baltic countries (Estonia, Latvia and Lithuania) and five are in Eastern Europe (Czech Republic, Hungary, Poland, Slovakia and Slovenia). The three Baltic countries and the five Eastern European countries have similar histories, in that from the end of the Second World War until the fall of the 'Iron Curtain', they had centrally-planned socialist economies and were isolated from contact with non-Soviet bloc countries (Figure 1.1). Over the last 15 years these countries have seen a gradual opening up of their borders to the west but in many cases greater restrictions on movement across eastern borders. Identity narratives in virtually all Eastern European countries frame the eastern border of a particular state as the eastern border of Europe (Kuus, 2004). In this way they seek to shift the discursive border between Western Europe and Eastern Europe further east and to thereby move themselves figuratively into Western Europe. This practice has been defined as nesting orientalism (Bakic-Hayden, 1995) and captures the flexibility of the Western/Eastern European framework. These changes will accelerate with accession to the European Union.

From May 2004 the frontier of the European Union has moved eastwards and this new border is one of exclusion at which would be immigrants are controlled. In the west movement across the border has been relatively easy for citizens of the European Union and of the neighbouring countries for several years. Although migration of citizens of the new EU members to some of the richer countries of the old EU is being restricted for a maximum of seven years, and the Schengen Treaty, which has enabled virtually free movement within the boundaries of the pre-May 2004 European Union, will not be applied immediately to the new members, the border changes are enormous and far reaching. On the Hungarian border with Romania, now the edge of 'Fortress Europe', border stations have already been brought up to EU standards, and exchange of information is helping to control people trafficking and the movement of drugs and stolen cars (Pál and Nagy, 2003).

In this volume we examine the impact of these changes through a case study of the eastern and western borders of Hungary with particular reference to gender and entrepreneurship. Small businesses operating in the peripheral regions of the European Union are 'an integral part of the rural space and the major alternative to agricultural employment, and thus creating and supporting rural businesses is considered a primary strategy for the survival of these areas' (Skuras et al., 2005). Such a strategy has not really gained momentum in rural Hungary. Although self-employment and entrepreneurship were seen as solutions to unemployment after the transition in Hungary, the specific problems of the rural areas have received little attention. In

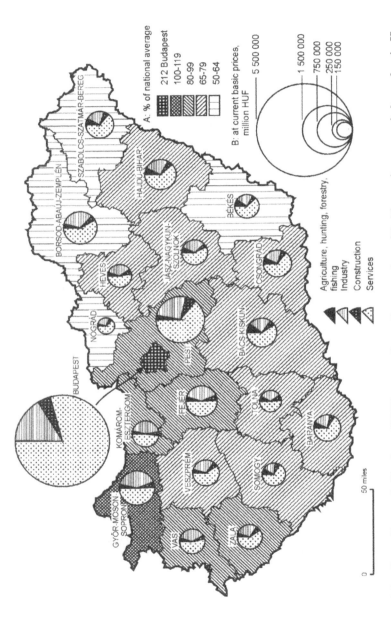

Figure 1.1   Gross Domestic Product (GDP) per capita and gross value added by main economic branches in Hungary by counties, 2002

Source: Regional Statistical Yearbook 1996, Hungarian Central Statistical Office, Budapest, 1997.
Regional Statistical Yearbook 2002, Hungarian Central Statistical Office, Budapest, 2003.

addition, the particular nature of post-communist entrepreneurship and its gender aspects have rarely been considered (UNECE, 2002).

Borders can be unifiers in terms of economy, population mobility and cultural ties but they can also be dividers of ethnic minorities and national identities. Borderlands are barometers of political change. Conditions of borderlands will vary according to the historical political and geographical context. There will also be differences in the experience of frontiers in the daily life of the population living in border regions. These social experiences are particularly affected by trans-border migrants in terms of size of such migrant flows and the type of migrants involved. Geographical scale makes a difference in cross-border cooperation as local links may be opposed by the state as happened on the eastern Hungarian border during socialist times. Today, within the European Union, there is a ranking of borders from internal open borders operating under the Schengen Treaty to those borders on the periphery of the EU which are crucial to the maintenance of 'Fortress Europe'. In this study we deal with two Hungarian border areas which differ on most of these characteristics.

Only 15 years before EU accession, Hungary had ended its membership in Comecon and the Soviet bloc. These two transitions in such a short period, from socialism to capitalist neoliberalism and from membership in an eastern Communist political association to becoming part of NATO and Western Europe, have had far reaching effects on the economy and people of Hungary. In this volume we examine the impact of the post-socialist transition and the preparation for joining Western Europe on villages in western and eastern border zones. This major restructuring of the Hungarian economy has demanded enormous social and economic adjustments by individuals. Most studies of these changes have focused on cities or on the nation as a whole. Here we look at rural areas and consider these adjustments at the household level. We deconstruct the household to look at the impact on gender relations and on women and men individually.

These changes have increased the level of uneven development in rural areas of Hungary. Data suggest that in Hungary women's economic activity rates show greater regional differences than men's, and labour market opportunities are also gendered, with women's being more determined by the geographical location of the region where they live than are men's (Timar, 2002). This place-based conceptualization of economic diversity and gender frames the following discussion of border regions, post-transition rural restructuring and the role of gender.

The Gendering of the Transition

The immediate impact of the post-1989 transition in Eastern Europe on the position of women attracted considerable interest from outside the region (United Nations, 1992; USAID, 1991; Watson, 1993; Einhorn, 1993a and b; Goven, 1993) but internally there has been resistance to women's movements in many countries (Adamik, 1993; Kiss, 1991; Répássy, 1991; Szalai, 1991) and geographers have especially lagged in interpreting economic and political change through a gender lens (Ciechocińska, 1993; Timar, 1993a and b). Inextricably linked to the transition to democracy was the ending of state control of the economy and a new encouragement of private enterprise, leading to fundamental change in the relationship between the state, the enterprise and the individual (Barr, 1994) with generally detrimental effects on rural areas and on women. Indeed changes in the position of women were an intrinsic part of this transition in Eastern Europe (Watson, 1993; UN, 1992; Kulcsár and Brown, 2000).

Although by 1998 all 27 post-communist countries of the region had signed and ratified the UN Convention on the Elimination of All Forms of Discrimination against Women (CEDAW), their participation was largely ceremonial. Human rights were not seen as individual rights but as collective rights, best expressed through the state. The state had need of labour so it provided education and health services universally and child care services so that women were not prevented by family responsibilities from carrying out full-time jobs. Unemployment did not exist. 'The underlying process was authoritarian rather than rights-based: a semblance of equity which often came closer to uniformity than genuine equality was imposed from the top down. Human rights and fundamental freedoms were denied, civil society was suppressed, and the family was neglected or viewed with suspicion' (UNICEF, 1999, p. viii). The communist state enshrined equality in legislation but did not allow it to flourish at the grassroots. Even today feminism has a negative image in Hungary and few Hungarian women dare to call themselves feminist. The thinness of the veneer of equality is seen in the collapse of the representation of women in Parliament which fell from one-third in the 1980s to 8 per cent today in Hungary. 'By shattering the state monopoly over economic, social and political issues, the transition has exposed women to an environment in which the conditions for equality have yet to be explored' (UNICEF, 1999, p. viii).

Kulcsár and Brown (2000) note that a revival of patriarchy has taken place in all post-socialist nations. This is especially seen in terms of family relations, reproductive decisions, employment and income, political representation and participation in civil society. However, despite the commonality of political organization, economic structure and ideology under communism, the transition to a market economy has strengthened differences in women's status and economic vulnerability both between and within countries (Szalai, 1991).

The economic restructuring following the end of the Cold War was embedded in the simultaneous process of social and political transformation (Quack and Maier, 1994) and was not gender neutral (Einhorn, 1993a). Watson (1993) argues that the transformation of the relationship between public and private spheres lies at the heart of the process of change in Eastern Europe, and that the exclusion of women and the degrading of feminine identity currently in train are not contingent to, but rather a fundamentally constitutive feature of, the democratization of Eastern Europe. The transition from a command to a market economy and from a centrally planned regime to democracy was accompanied by a decline in state-provided services, an increase in pornography and prostitution, pressure from politicians and church to limit access to abortion (Gal, 1994), and increased public support for traditional male and female stereotypes (Unicef, 1999; Makara, 1992; Goven, 1993; Matynia, 1995).

Under socialism, citizenship was linked to employment and both women and men were expected to be in full time paid jobs (Timar, 1993b). Thus work became a prerequisite for access to social services and indeed 'eligibility rights based on citizenship were substituted by ones based on having regular and continuous employment' (Szalai, 1991, p. 153; Ferge, 1992). Despite this apparent official support for gender equality (Bollobás, 1993), women's double burden meant that in terms of the number of hours worked per week the gender gap was greater in Eastern Europe than in any other industrialized region of the world (Einhorn, 1993a). Table 1.1 shows that working time for both men and women in Hungary has decreased in the last 30 years but women continue to work longer hours than men. The overall decline in hours spent on reproductive work since 1963 reflects improvement in living conditions and the spread of labour saving devices but the increase between 1986 and 1993 is probably a response to the loss of state supported childcare facilities.

Table 1.1   The gender division of time spent annually in production and reproduction in Hungary, 1963–1993

| Year | Production | | | Reproduction | | | |
	Women's hours total and (%)	Men's hours total and (%)	Total hours worked	Women's hours total and (%)	Men's hours total and (%)	Total hours worked	Grand Total of hours worked
1963	71 (32.7)	146 (67.3)	217	134 (81.2)	31 (18.8)	165	382
1977	90 (40.4)	133 (59.6)	223	107 (74.3)	37 (25.7)	144	367
1986	83 (39.2)	129 (60.8)	212	103 (74.1)	36 (25.9)	139	351
1993	57 (36.1)	101 (63.9)	158	105 (72.4)	40 (27.6)	145	303

Source: Falussy, Béla and György Vukovich (1996), 'Az idő mérlegén (On the balance of time), (1963–1993)', in R. Andorka, T. Kolosi and Gy. Vukovic (eds), *Társadalmi Riport, 1996 (Social Report, 1996)*, Budapest: Tárki Századvég, pp. 70–103.

In 1989–91 women made up between 45 per cent and 51 per cent of the workforce of Eastern Europe and 70 per cent of women of working age were either employed, on maternity leave, or undergoing training (Einhorn, 1993a). In the early years of the socialist economy women were needed as workers because of the loss of a high proportion of men during the Second World War. As most men did not earn enough to support a family, a second income became a necessity. Following the first postwar year of sub-replacement fertility in 1958, the Hungarian government established a number of programmes explicitly aimed at raising fertility (Barta et al., 1984; McIntyre, 1985). Similar policies were followed elsewhere in Eastern and Central Europe with Romania utilizing the most extreme measures (Momsen, 2004). The State provided a broad array of assistance, such as maternity leave, family allowances (Jarvis and

Mickelwright, 1992; Mieczkowski, 1985; Kulcsár, 1985), and crèche facilities, to encourage women's reproductive role. Women in Hungary sometimes used maternity leave to increase their involvement in the 'second economy', that is outside the state sector, and/or to improve their educational qualifications (Vásáry, 1987).

Hungarian women and men have equal access to education and recently women have outnumbered men in tertiary education. However, this has not translated into higher salaries as blue collar jobs held by men, such as skilled work in heavy industries, were generally paid more than white collar service jobs held by women. Women have broken into fields that were traditionally male such as medicine and engineering, but gender segregation of employment is as marked as in Western Europe (Einhorn, 1993a) despite state socialism's gender-blind agenda. Under socialism, women were much less likely than men to move from unskilled to skilled or professional work and less likely to be able to move out of agricultural occupations. The official retirement age for women workers was 55, lower than that for men, and it has been suggested that this policy related to the important role for grandmothers as supplemental childcare providers (Maltby, 1994).[1] It is clear that gender equality was more ideological and legal than real in Eastern Europe under a socialist, patriarchal-paternalistic system (Völgyes and Völgyes, 1977; de Silva, 1993).

There have been very few studies of regional differences in the impact of the transition on gender at the sub-national level. Unicef (1999) published a study of *Women in Transition* covering all the post-socialist countries at the national level and Katalin Kovács and Mónika Váradi (2000) have examined the impact of the transition on women in a small central Hungarian town. Studies of changing gender roles in rural areas are especially rare with Bettina van Hoven-Iganski's (2000) study of rural women in the former East Germany being one of the few.

Uneven Development

Hudson (2003) argues that the changes in Europe following the collapse of state socialism in the Eastern European countries have become both complex and often contradictory, 'simultaneously creating processes of convergence and divergence, homogenization and differentiation' (Hudson, 2003, p. 49). The application of EU common regulatory processes has encouraged homogenization but at the same time the introduction of neo-liberal market forces in the Central and Eastern European countries has increased both individual and regional differences in wealth. These changes have also affected regional labour markets both qualitatively and quantitatively. However, as Smith and Ferenčíková (1998) point out, it is only recently that attention has turned to the sub-national regional development implications of transition in East and Central Europe. Figure 1.1 shows the different sources of per capita Gross Domestic Product (GDP) at the county level. It can be seen that per capita GDP in 2002 was much higher in the west of the country than in the east. Budapest has the highest per capita GDP but the only other place with an above average per capita GDP was the county of Győr-Moson-Sopron. Eastern counties, including Békés County, had average per capita GDPs barely more than half the national average in 2002. Services and

_____

[1] Retirement age is now the same for men and women at age sixty-two.

industry are the most important sources of GDP in Győr while in Békés services and agriculture are the leading sectors. Figure 1.1 shows clearly the marked regional differences in terms of sources and levels of per capita GDP. We hypothesize that these regional differences will be reflected in differences in entrepreneurial success.

Few regional studies in Hungary (Enyedi, 1990 and 1994; Cséfalvay, 1994; Hamilton, 1999; Horváth, 1999; Momsen, 2000) consider the networks and social practices vital to entrepreneurship which are clearly territorially embedded (Smith and Swain, 1998). In order to illuminate regional differences we selected one study area on the northwest border with Austria in the county of Győr-Moson-Sopron and one in the southeast in Békés County on the Romanian border. The two areas were chosen to maximize the range of conditions under which rural entrepreneurs operate: the north-western border with Austria is an area of intensive mixed farming and a dense network of nucleated settlements which benefits from flows of capital, ideas and tourists from Western Europe and daily movement of Hungarian workers across the border to better paying jobs in Austria; while in the more thinly-populated southeast where there is extensive grain and livestock production, cross-border contacts with Romania are predominantly through illegal immigrants and black-market traders seeking opportunities in Hungary. Pál and Nagy (2003) suggest that the post-transition economic changes in the eastern frontier area have been negative, pointing out that the regional rural workforce went from working in the black economy to the grey and now to being unemployed. Despite these differences, in both areas rural entrepreneurs are responding to demand for services from local residents and from the externally generated growth of travellers and tourists. To add an international dimension we also looked at the situation of entrepreneurs on the Romanian side of the eastern frontier area.

This comparison allows us to take into consideration the effect of the flow of capitalist influences from west to east. The county of Győr-Moson-Sopron was traditionally an area of feudal estates and peasant farmers while Békés county in the east was an area of large grain farms worked by an agricultural proletariat. In the Győr area it was predicted that cross-border trade and flows of capital and tourists would encourage the growth of entrepreneurship among a population historically attuned to self-employment, while in Békés County the presence of the border with Romania was expected to have a neutral effect on the growth of small firms. The western study area is described by Enyedi (1994) as an area with promising prospects while he sees the eastern area as a crisis region. The northwest has fewer foreign residents than the southeast but both have net outmigration (Geographical Research Institute, 1994 and 1995). There is considerable foreign investment in the west but very little in the east (Berényi, 1992) and the rural economy of Győr-Moson-Sopron is more diversified than in Békés County. Thus the conditions for the development of entrepreneurship are very different in the two regions.

In our study we tried to maximize the sub-national regional differences by looking at gender differences in two very different border areas, with a focus on entrepreneurial activity. The study areas were selected to represent the range of conditions under which Hungarian borderland rural entrepreneurs operate: with one area in north-west Hungary on the border with Austria which has benefited from proximity to Western Europe, and the other area in south-east Hungary on the Romanian border where cross-border contacts are limited. On the western border there were already joint planning projects in the 1980s with a cross-border national park and after 1995, when Austria joined the European Union, projects funded by the European Union through the INTERREG (initiatives promoting cross-border cooperation in the European Union's internal and external border regions) and PHARE-CBC (programme supporting cross-border cooperation at the external borders of the European Union) were set up along both sides of the western border. In the east there was no such early trans-border collaboration and contacts have been limited. However, religious pilgrimages across the border into Romania are beginning to be revived and in March 1997 a regional agreement for cooperation was signed by the border villages of Méhkerék and Nagyszalonta, with the aim of establishing a joint enterprise for growing vegetables (Pál and Nagy, 2003). For many Romanians legal cross-border visits are still limited by the high cost of passports.

Inextricably linked to the transition to democracy was the ending of State control of the economy and the encouragement of private enterprise, with generally detrimental effects for both women and rural areas. These changes introduced an element of choice into the lives of women and men by increasing the range of possibilities and opportunities for livelihood strategies which differed between urban and rural areas. This freedom to choose is essential to development and is 'the principal determinant of individual initiative and social effectiveness' (Sen, 1999, p. 18). The exercise of freedom and the individual agency it encourages are mediated by values which may also affect the presence or absence of corruption and the extent of trust within a community. The speed at which the post-socialist transition took place led to a new spatial variation in local values. These differences are related to varying levels of isolation from both major Hungarian urban centres and from external influences. The two border regions studied exhibit quite different attitudes to the transition and so of changes in local values and social capital.

In our study we identified the factors which assist in the development of entrepreneurship in rural areas and those which act as constraints. In particular, gender differences in types of small businesses and in enterprise survival rates were examined and the impact of entrepreneurial activity on gender roles and relations considered. Regional and gender differences in social capital, as it effects cross-border business activity, were assessed. Spatial differences in the location of small businesses within the community and region were analysed and considered in relation to household relationships. Family and community attitudes to entrepreneurship and changing gender roles in the two regions are compared.

As noted by Labrianidis (2004, p. 16) '[E]ven the most remote rural areas are becoming more and more integrated into wider spaces of interdependency, leading us to consider globalization processes and the evolution of the EU in particular, as crucial contextual factors in our analysis'. Following Labrianidis, we compare the effect of cross-border influences, as a surrogate for capitalism and the spread of the global economy, on the growth of entrepreneurship in our two rural study areas.

Research Methodology

The project used mixed mode methods of data collection so that triangulation of information could be achieved. These methods included both extensive large scale questionnaire surveys and intensive methods such as key informant interviews, unstructured in depth surveys, life histories and focus groups. In addition, local village-level records of registered entrepreneurs dating back several years were consulted. Secondary sources such as county-level employment records and population statistics were also utilized and maps produced by the National Geographical Institute were consulted.

On the basis of the most recent village lists of officially registered entrepreneurs, a random sample was selected. Questionnaires were developed and field-tested covering demographic information, types of businesses, problems and constraints for entrepreneurial activity, family background, external linkages and reasons for deciding to become self-employed. This information was gathered through questionnaire surveys of 351 individual male and female entrepreneurs in 17 western Hungarian border villages and ten border villages in south-eastern Hungary. The informants were stratified by gender and economic sector and the surveys carried out by Hungarian university students. These questionnaires covered a 30 per cent random sample of the registered entrepreneurs in the two border areas in 1998. These registers provide data on types of business activity, the name of the entrepreneur and sometimes the date of the foundation of the business. Where the date of establishment was available we were able to trace the survival rates over time.

The quantitative analysis of the questionnaire surveys is supplemented with qualitative data. All three authors participated in unstructured interviews with key informants in most of the villages. These key informants were usually the mayor or another member of the village administration. They were able to give us an overview of the role of entrepreneurship in the economies of their own villages and describe some of the particular local problems faced by their villages in relation to the border and the presence or absence of border crossings. Focus groups of women entrepreneurs, funded by the Gender Research Centre of the European University, were held in the east and the west. To encourage participation travel expenses were refunded, refreshments provided and participants were offered an opportunity to meet with a woman psychologist from Budapest. Unfortunately despite previous commitment, few women came for the focus groups. Most pleaded shortage of time but it is thought that it was the lack of trust and openness between individuals in post-socialist Hungary that was the main barrier, at least for some. Neo-liberal policies have brought about a growing polarization of society and the intrinsic competitive nature of entrepreneurship has led to a dependence on family and the rupture of relations with neighbours.

Personal narratives or life histories were collected from a small sample of the men and women entrepreneurs in both regions. Such life histories have been increasingly used by feminist researchers (Sizoo, 1997; van Hoven-Iganski, 2000). Miles and Crush (1993, p. 85) summarize the advantages of personal narrative techniques as follows:

First, they are seen as corrective to the silences . . . of many archival documents . . . Second, they may help the researcher to achieve a degree of depth, flexibility, richness, and vitality often lacking in conventional questionnaire-based interviews . . . Third, they may help uncover not only what people did, but what they wanted to do, what they believed they were doing, and what they now think they did. Fourth, they can illuminate both the logic of individual courses of action and the effects of systemic and structural constraints within which life-courses evolve . . . And finally, they may assist to crack the codes of the muted group by documenting the personal histories and struggles of 'invisible' people, thus giving voice to the marginalized and the dispossessed.

Feminist geographers have tended to focus on qualitative methods rather than quantitative but Staeheli and Lawson (1994) pointed out that feminist fieldwork may also involve archival research and extensive surveys. Attempts to consider a gender perspective in entrepreneurship studies have mostly been anchored in theories of entrepreneurship rather than in feminist theory. Steyaert and Bouwen (1997) suggest that the reason that personal narrative analysis is rare in entrepreneurship studies is related to the field's epistemological foundation. However, this is changing as feminist researchers expand their interest in this field (Berg, 1997; Carter, 1993; Oberhauser, 2004; and Wheelock et al., 1999). Personal narratives in our case helped us to discover aspects of an individual's history which had encouraged an interest in starting a small business and gave us some insight into the impact of changes in women's roles on family gender relations. However, personal narratives are not unproblematic sources for the construction of experiences and events (Miles and Crush, 1993). They aim to give voice to the subaltern but in so doing the positionality of the researcher has an influence that is not always easy to assess (Moss, 2002).

Our data on life histories were collected by the two Hungarian authors but in many cases, as in the key informant interviews, the first author also participated. The presence of an individual who has little facility with Hungarian had an interesting effect. In general, the interviewees were pleased that someone had come all the way from California to hear about their lives and problems. At the same time the foreigner was also identified as a Western European familiar with the European Union and its rules. Many of our respondents were not sure whether the impending incorporation of their country into the EU was a threat or an opportunity and so were interested in her views on this matter. As an outsider she was non-threatening as she clearly did not have any links to the Hungarian authorities and was not capable of gossiping with their neighbours. At the same time her experience of interviewing in many different cultures gave her an awareness of body language and tone of response that were sometimes missed by someone concentrating on the verbal information being provided. The outsider was also able to detect non-sequitors and lacunae in the narrative and could interject additional questions to fill these gaps. The interviews were held in people's houses and their surroundings also raised additional questions that might not have been noted by someone who was more familiar with the local environment. Overall, we found the combination of insider/outsider participation to be a positive experience for both the interviewers and the interviewee. However, as feminist geographers working in a country in which feminism is a new and somewhat threatening concept, and where the Hungarian Academy of Sciences has declared that there are no gender issues in Hungary, we found ourselves in a somewhat paradoxical space (Moss, 2002) during our fieldwork.

Data gathering was undertaken over the period 1997–2001. An additional visit to both field areas was made by the first author in September 2003 to look at recent changes. The research was funded by the National Science Foundation of the USA (Award number SBR-9710073). The Fulbright Foundation, through a Fellowship, funded a six-month visit to the University of California, Davis for Judit Timar in 2000 and Irén Szörényi also spent a month in Davis that year with NSF funding. These visits facilitated joint analysis of the research data. The original concept for the research was developed during a visit by Irén Szörényi to California in 1993, funded by the Hungarian Academy of Sciences, and further refined during visits to Hungary by Janet Momsen in 1994 and 1996, funded in part by the Hungarian Academy of Sciences. Field periods in Hungary in 1997, 1999 and 2001 were undertaken by Janet Momsen working with Judit Timar and Irén Szörényi. The NSF also funded a Romanian-speaking University of California, Davis undergraduate, Margareta Lelea, to study entrepreneurship on the Romanian side of the border in 1999. This project has been an enjoyable collaboration and we are very grateful to our various funders.

Structure of the Book

The book combines a review of previous literature with analysis of original data collected in 17 border villages in east and west Hungary. It falls into two parts: an introduction followed by three chapters reviewing the literature on various aspects of the topic; in the second part there are three chapters reporting on the findings of the survey and we end with a summary conclusion. Following the introduction, the second chapter looks at the role of borderlands in Hungarian history and traces the changes in the nation's borders over the last millennium. Recent changes resulting from accession to the European Union and the attitudes of entrepreneurs to these changes are considered, especially in their differential effect in east and west Hungary and the growing importance of an active rather than a passive transnationalism in the border areas under study. The third chapter looks at rural areas in Hungary, and considers the nature of rural peripherality, especially as it applies to borderlands. It considers the situation of rural lands both before and after the transition, looking in particular at the impact of land privatization on entrepreneurship. Rural-urban differences in employment, incomes and time budgets are also analysed. Rural gender divisions of labour are considered over time in relation to the impact of changing structures of land ownership. Demographic changes in rural areas are also looked at especially in relation to suburbanization and counterurbanization. The fourth chapter reviews some of the recent studies of rural entrepreneurship in Europe and compares their findings to the situation in Hungary. This chapter also presents a literature review of the historical development of Hungarian entrepreneurship and the relative importance of men and women entrepreneurs. It emphasizes the importance of the pre-1989 private economy in laying the foundation of entrepreneurial activity in post-socialist Hungary.

In the second section of the book we introduce the results of our research. Chapter 5 describes the general characteristics of the entrepreneurs surveyed in the two areas. Chapter 6 looks particularly at differences in social capital among the self-employed in the two areas and presents the results of a Principal Components Analysis of the survey data. This analysis includes both gender and regional differences. The next chapter reports on the impact of entrepreneurship on women and on gender relations within the household and community and is based on in-depth interviews with a few of the more successful self-employed men and women. Our final chapter presents a synthesis of our findings about entrepreneurship in the two border areas studied and their contribution to rural development. A survey of border villages in western Romania reinforces these comments.

Chapter 2

Borderlands

Borderlands acquire their basic identity from interaction with the frontier and its rules and from cross-border transactions, but are not just passive spaces (Morehouse and Pavlakovich-Kochi, 2004). They function as a space 'in-between' where 'existing understanding and knowledge of the world can be deconstructed, examined, perhaps changed, then reconstructed' (Morehouse, 2004, pp. 30–31). Furthermore, Clement (2004) argues that globalization has led to border regions being increasingly seen less as barriers to trade and more as contact zones between neighbouring countries encouraging the development of cross-border regional cooperation. The Iron Curtain was an archetypal barrier imposed by global super-power ideologies rather than a frontier between European countries. The post-communist changes in Eastern Europe have both encouraged new cross-border relationships and revived older frictions between neighbouring states.

We chose to focus on border areas because the frontier situation highlights the problems of the post-communist economic transition and because borders are so important in the history and geography of the country. Hungary's borders were relatively stable from the establishment of the kingdom a thousand years ago in 996 until the end of the First World War. 'This permanence is still embedded in Hungary's collective consciousness: the concept of the "Thousand Year Border" has consequently formed an important part of national policy' (Hajdù, 1996, p. 139). By contrast, the most characteristic features of twentieth century Hungarian history have been frequent and radical border changes. For this reason, border-related issues have become a crucial element of Hungarian domestic and foreign policy, as well as of the nation's social and political geography.

A century ago Hungary was seen as a 'bastion of Western Europe projecting eastwards: the country indeed for ten centuries of European history, played the part of a breakwater against which the waves of Oriental barbarism dashed and were broken' (de Vargha, 1910, p. 1). With its accession to the European Union, Hungary once again assumes its 'breakwater' role, though this time against the perceived threat from the Eastern hordes of illegal immigrants.

In a sense, the whole country can be considered a 'borderlands society' (Hajdù, 1996, p. 139). Hungary borders seven countries: Austria for 356 kms, Slovenia 102 kms, Croatia 355 kms, Yugoslavia 164 kms, Romania 453 kms, Ukraine 137 kms and Slovakia 679 kms (Rechnitzer, 2000). The Trianon Treaty at the end of World War I dispersed approximately two-thirds of Hungary among its neighbours so that today over three million ethnic Hungarians live outside the present borders of the country. Currently, 35 per cent of its territory can be considered a border region and 28.2 per cent of the population, about 2.7 million people, live in such regions (Ibid). Nearly 30 per cent of Hungarian towns are located close to the border. Out of Hungary's 19 counties, only five do not share borders with a neighbouring state; [and] 'about

60 per cent of Hungary's borders (including four-fifths of the Hungarian/Romanian) sever linguistically and culturally cohesive regions' (Hajdù, 1996, p. 140) According to Hajdù (1996), only the Hungarian-Austrian state border coincides entirely with natural linguistic borders.

There are 312 villages and towns – about 10 per cent of all Hungarian settlements – located directly along the 2,242 kilometres of the national border. These are considered in a sociological sense as 'frontier settlements'. The main social and economic characteristics of these settlements are determined by the border with many being cut off from much of their natural hinterlands by frontier re-alignments (Kovács, 1989; Kovács, 1993; Böhm, 1995). In the study reported on here we looked at 17 villages on the Hungarian border with Austria and ten on the border with Romania. In this latter region villages have suffered most from being isolated from their former hinterlands and by political reinforcement of the border severance of a linguistic region.

By definition border regions are peripheral and this was especially marked in the era of state socialism in Hungary. The concept of periphery has developed different meanings over the last few decades: periphery as distance, as dependency, as difference and as discourse (Ferrão and Lopes, 2004). The first two meanings tend to have negative connotations in terms of lack of accessibility and relationships to a distant and more prosperous core and are clearly related to uneven spatial development. However, the growth of local responses to global restructuring has allowed the emergence of some success stories for peripheral areas. This has led to the concept of periphery as difference, with interest in local foods, 'terroir' and unspoilt environments giving peripherality a positive aspect. So it can be argued that border areas as peripheral zones have the potential to grow or decline depending on finding new markets, new industries such as tourism and manufacturing on greenfield sites which benefit from the attractive environment and available space, and levels of entrepreneurial innovation (Muller et al., 2002), among other things.

The relative backwardness of Hungary's border regions, compared to core areas, differs from border region to border region influenced by historical events, as well as changes in the political and economic system. In the era of state socialism the centralized state economic control model prevented local cross-border contacts, even with 'fraternal countries' who were also members of the socialist bloc. The few cross-border contacts that did operate occurred when the leaders of a county wishing for such a contact submitted the request to the central authority and the central government of both countries agreed. These links always depended basically on the relation between the two countries on either side of the border and so were suspended or terminated frequently (Rechnitzer, 2000).

Hungarian military geography theoretically has long considered a 25 km wide strip along the state border as a topographical and a 50 km wide strip as a strategic borderland area, defining both as danger zones. Hajdù (1996, pp. 140–141) states that 'Hungarian settlement and borderland development policy and practice were influenced by this theory.' Within COMECON the issue of common development initiatives for the borderland areas of member states was first raised in 1960 and was focused on a problem-oriented approach to border regions. 'No real attempt was made to promote or even emphasize the importance of day-to-day contacts within international border regions' (Hajdù, 1996, p. 142).

International cooperation between regions is very important for border zones of Hungary (Tóth, 1998). This cooperation can take place in various ways but the main methods are as follows: cooperation between individual settlements on either side of an international border; international cooperation based on a regional organization formed between administrative units on the border of two or more countries; and inter-regional cooperation between units at different levels without the borders having any significant role in dividing the region (Ibid). In most cases connections, both internal and external, are deformed in Hungarian border zones but the differences are greatest between the western and eastern zones. Settlements in western Hungary are more open to the relatively more developed west of Europe and despite some economic problems, Tóth (1998, p. 70) feels that they are more able to 'internalize advanced Western technology, management structures, innovation and capital'.

The Austro-Hungarian Border

At the beginning of the twentieth century the Austro-Hungarian border did not exist as the whole area was part of the Austro-Hungarian Empire. Following the collapse of the Empire after the First World War the linguistic boundary became the border between the two countries. In 1945 the Austrian territories adjacent to Hungary were occupied by the Soviets and so the Hungarian-Austrian borderlands were controlled by the same foreign power. When the communists took over power in Hungary in 1948–49, Hungary's relations with Austria deteriorated, adversely affecting the border areas. After the Austrian Treaty of 1955, the occupying Soviet troops were withdrawn and Austria regained its sovereignty and became an independent but neutral country. Its 365 km common border with socialist Hungary became an almost completely impenetrable frontier between the two worlds. The 'Iron Curtain' came down and all along the border minefields were laid (Plate 2.1). From 1950 until 1969 there was a strict border zone 15 kilometres wide which people could not enter or leave without permission. The border villages were closed and soldiers boarded trains going towards the border to check identity papers. Soldiers were stationed in these villages to man guard posts along the border. Many people moved out of the border area because of the restrictions. The first cracks appeared briefly during the Hungarian revolution in 1956 when more than 200,000 people emigrated across this border. From 1 May 1969 movement to the border became easier and the restricted border zone was reduced to two kilometres in width and the number of closed villages reduced.

In the 1970s there was more legal movement within the border zone and more interaction across the borders. In the west the isolation of the Hungarian border villages within their own country encouraged interaction with Austria in terms of people smuggling and, as television sets became available, watching Austrian television. During the 1980s the frontier gradually became more permeable and less restrictive and border areas began to be popular areas for contacts between Hungarian families, and those from other Eastern European countries, who had been divided by the Iron Curtain. Austrians also discovered Hungary as a source of cheap services and restaurants. Research on border region problems and transborder regional planning began in 1967 but stopped in 1972 because of lack of interest by Austria. It started again in the 1980s with the initiative coming from the Austrian side generated by the

Plate 2.1   A relic piece of the 'Iron Curtain' on the western border

growth of large scale transborder shopping and tourism. By 1985 this border had become the busiest between East and West. Cross border commuting also intensified, first in the form of contract labour and later in terms of large flows of inexpensive illegal Hungarian labour. Thus the border in the west became more of an advantage economically and socially than the disadvantage it had been earlier. On 2 May 1989 the electric fence along the border was pulled down. On 1 June 1989 the restricted border zone and the closed village designation officially ended.

These new advantages of this border zone were as follows:

- There were no areas of post-socialist economic crisis since there was no heavy industry in the area.
- Consequently there was very little industrial pollution.
- The long isolation of the border had reduced the population and protected the environment so the bi-national Fertő-Hanság National Park could be established with its unique flora and fauna and landscape heritage (Kukorelli Sz. et al., 2000).
- The work force in the region is well qualified and adaptable because of its experience of cross-border guest work (Ibid).
- The open border and a good telecommunications network allow for the adoption of innovations.
- The presence of a German minority population has encouraged the language ability of the border population.
- Austrian capital is invested in new factories in the border areas, often on greenfield sites, and disseminates inland (interview with mayor of Jánossomorja, June 1998).
- Low unemployment levels partly because of new investment and of job opportunities on the Austrian side of the border (Ibid).

The most advantageous border locations are those of the settlements with border stations and the villages with German minority populations. Sopron and its surroundings, the settlements around the Fertő Lake, and the major crossing points of Jánossomorja and Hegyeshalom can thank their improved economic positions to the proximity of the Austrian border. However, the open border also brought disadvantages related to lax security in the early years. Prostitution, drug dealing and the smuggling of people and goods spread. Special police and military forces on the Austrian side of the border controlled the border but economic migrants caught trying to enter Western Europe were the responsibility of Hungary and the border regions had to provide temporary accommodation for these illegal migrants (Kukorelli Sz. et al., 2000). This responsibility has now moved to Romania on the eastern border of Hungary.

Cross Border Regional Cooperation

The counties closest to the Austrian border were the first to become members of the Alps-Adriatic Working Community, which fostered interregional cooperation. In 1985 a joint planning council, the Austro-Hungarian Regional Development and Physical Planning Commission, was created by Burgenland and Győr-Moson-Sopron for collaboration in planning and regional development. The aim of the Commission was mutual information exchange about development in the border zone and the harmonization of planning on both sides of the border. This marked the birth of border cooperation and planning but still with leadership from the central government (Rechnitzer, 2000). Meetings were held in 1986, 1988 and 1993 at which the collaboration was confirmed and sub-committees were set up. Concrete plans were made for the joint development of the Sopron-Kőszeg tourist district and of the Ferto Lake area. From 1994 the Commission became the PHARE-CBC committee. In 1992 the Border Regional Council was established on the initiative of two Hungarian counties (Győr-Moson-Sopron and Vas) and Austrian Burgenland and they drew up recommendations for cooperation. The main task of the Council is cooperation in the fields of transport, telecommunications, energy supply, environmental protection, economy and education. It has planned for waste disposal sites and encouraged the opening of several border stations. The European Union subsidy began in 1995 through the PHARE-CBC programme and gave considerable impetus to border cooperation. In 1998 the Interreg PHARE-CBC programme received 35 million ECUs to strengthen collaboration and decrease differences in lifestyles and quality of life on either side of the border. It also aimed to rebuild economic and social collaboration and to harmonize planning systems with European Union institutions. Today intensive economic, social and local political interaction is occurring. The National Regional Development Concept, adopted by the Hungarian Parliament in March 1998, also encourages cross-border cooperation and relations and, thereby, contributes to the better use of potential regional centres that have become peripheries of the country through the creation of political boundaries (Horváth, 1999b). Funding is directed towards assisting cross-border cooperation between border regions, common planning and coordinated development.

Intensive economic, social and local political interaction is developing. Cycle paths have been built throughout the region and are now heavily used by tourists. Under the auspices of the programme a port on the Danube has been opened and a container terminal at Sopron has been built. Environmental control of garbage has been established in order to protect the natural resources of the region. A daily maximum of 550 legal workers are allowed to commute from Győr-Moson-Sopron and Vas counties across the border into Burgenland in Austria. After the six-year mandate of the Border Regional Council expired in 1998, the members (Burgenland, Győr-Moson-Sopron and Vas counties) established the West/Nyugat Pannónia Euroregion. Zala County joined in 1999. Its main body is the Council of the Euroregion with 40 members and although it is an important step to have a continuously operating institution, it is only slowly becoming fully functional for transborder cooperation (Rechnitzer 2000). The joint Austrian/Hungarian Fertő-Hanság National Park is a major element in the cross-border collaboration with a magnificent new Park Headquarters in Hungary. However, local people tend to feel that they are not benefiting much from tourists attracted to the Park. They see the Park as an 'island' separate from the surrounding area and such attitudes lead to conflict, as is common in areas on the edge of National Parks generally.

The privatization of the Hungarian economy has attracted substantial flows of Austrian capital some of which is being used in illegal ways such as by the unlawful purchasing of farmland by Austrians although Hungarian law forbids the purchase of agricultural land by foreigners. This is done through use of Hungarian citizens as front owners with the Austrians hoping to assume full ownership when legal restrictions are lifted. Between the World Wars double ownership was quite common when many landowners had their properties split by the new international border. Today, many foreigners buy second homes in Hungarian villages. This development has led to higher house prices in many western border communities and is making it difficult for local young people to buy their own homes in these villages.

However, Austria's membership of the European Union from 1 January 1995 created new conditions since the border became a frontier separating Hungary from the European Union. Following the Schengen Agreement in April, 1995, which ended the requirement for border controls on most state frontiers within the European Union, frontier controls on the Austro-Hungarian border became stricter. The presence of additional border police did, however, create new economic opportunities for residents of border villages. On 1 May 2004 this border became an internal border within the European Union and transborder interaction became much easier, although the Schengen Agreement has not yet been applied to this border.

The Hungarian-Romanian Border

In this area the peripheral border zones meet peripheral regions. There are no significant differences between the regional resources. For example there is ample labour on both sides but there are not enough workplaces. There are agricultural products on the Hungarian side and natural resources on the Romanian side but trade does not occur because of lack of capital. There is a bottleneck in cross-border communications because of accessibility, the limited number of crossing points (see Figure 2.1) and the bureaucratic nature of negotiating the border. Finally the differing legal and institutional frameworks plus unstable local currencies and high rates of inflation, especially in Romania, curb cooperation (Rechnitzer 2000).

On the Romanian side of the border, the proximity of Hungary makes this region more favourable for business than other parts of the country. The four western counties of Romania, Arad, Caras-Severin, Hunedoara and Timis have over half of all the nation's businesses (ADAR, 1996). Arad County had 13.4 per cent of all the new businesses in Romania in the mid-1990s (Ibid). However, geriatrification is occurring most markedly in rural areas, to an even greater extent than in eastern Hungary, with the village of Variaşu Mic having over four-fifths of its population officially retired (Lelea, 2000). Infrastructure in the villages is far worse than in the Hungarian villages on the other side of the border, and roads between villages, where they exist, are often unpaved and so impassable in bad weather. Generally, in order to reach a neighbouring village it is necessary to travel into the main city of Arad and then out again because there is no road linking the villages (Lelea, 2000).

'According to official propaganda, the Hungarian-Romanian border should have been an area of friendship and cooperation between socialist "brothers". [However], because of differing attitudes towards the issue of minority rights during most of the period of state socialism, the Romanian-Hungarian border symbolized alienation and even downright hostility' (Hajdù 1996, p. 144). This border transects Hungarian linguistic areas which should have eased trans-boundary cooperation. In practice it became the main hindrance, as for the Romanian government it was more important

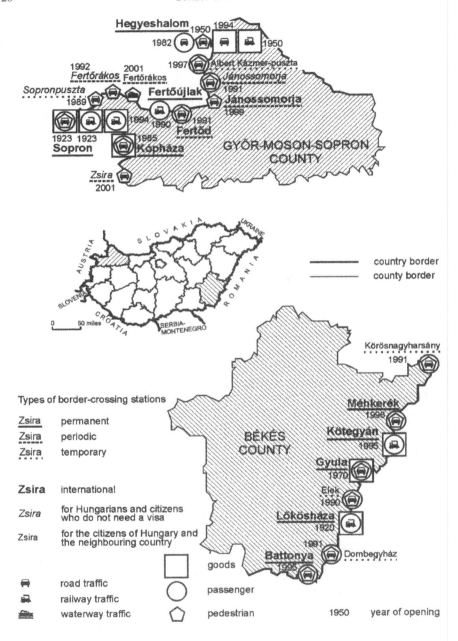

Figure 2.1   Border crossings in the Hungarian study areas

Source: Balcsók, I. and Dancs, L. (2003), and fieldwork.

to change the frontier's ethnic character than to encourage cross-border contacts. It is estimated that there are almost two million Hungarian speakers living in Romania. This is now considered a border of intensifying cooperation although the expectations of better interstate relations and more liberal border regional policies raised during the Romanian Revolution (1989) have only partly been fulfilled. However, the liberation of individuals has led to more trans-boundary movement than ever before. Much of this involves 'black market' trade, and illegal workers.

During the 1980s and early 1990s many Hungarians made regular trips into Romania with packages of food and clothing for relatives, and trans-border shopping by Hungarians occurred, stimulated by the price differentials between the two countries. The attitude of villagers to the Romanians who appear illegally in their midst is mixed, despite the commonality of language. Some villagers blame these illegal migrants for every theft that occurs, while others pity their poverty and offer them food and a few, never the actual informant, employ them cheaply and illegally during periods of high seasonal labour demand and 'because locals are lazy'. Only about 1 per cent of the population living in Hungary is ethnically Romanian but they have developed several local support organizations since 1989.

A perhaps unexpected benefit to the Romanian-speaking Hungarian villages in this region is that the central government during the communist regime provided a large amount of financial assistance to these communities in an effort to show that Hungary treated its 100,000 Romanian-speaking people better than Romania treated its Hungarian speakers. There are schools in which Hungarians of Romanian ethnicity can be taught in Romanian thus encouraging the continued use of the language. In addition, these areas are eligible for development assistance from the European Union because of their relative poverty within Hungary. The Hungarian-Romanian PHARE CBC programme began in 1996 with 35 per cent of the funds going to economic development, 30 per cent to culture and education, 17 per cent to environmental protection, 8 per cent to transport and 10 per cent to other fields. The Duna-Körös-Maros-Tisza Regional Co-operation area, a type three form of cooperation according to Tóth (1998), was set up in 1998. It involves six million people and includes the border areas of southeastern Hungary, Romania and Serbia. Its aim was to intensify cross-border contacts and to work out a uniform development plan for the region. Border cooperation is being encouraged in Romania with a programme called 'Activitate Transfrontaliera' (Transborder Activity). This programme, sponsored by the Open Society Foundation, enables the fourteen mayors in the frontier zone to meet twice a year (Kiss, 1999). A lower level of cooperation between two settlements was set up in 1997 between the villages of Méhkerék and Nagyszalonta with the aim of establishing a joint enterprise for growing vegetables (Pál and Nagy, 2003). Turmoil in the former Yugoslavia and the accession of Hungary to the European Union has made such common planning more difficult, but a duty free zone has been set up in Curtici in Romania, close to the Hungarian and Yugoslavian borders.

On 25 April, 2004, a few days before Hungary entered the European Union, a Reconciliation Park was opened in the Romanian town of Arad, close to the border. This Park commemorates the victims and heroes of the 1848 revolution in both countries and indicates a movement towards rapprochement between the two countries as Romania prepares to enter the EU in 2007 (*Herald Tribune*, 26 April 2004, p. 3). The then Prime Minister of Hungary, Péter Medgyessy, whose family originally came from Transylvania, hailed the geopolitical significance of the recent improvement in relations between the two countries at the inauguration of the Park. He said 'I think Europe will win through this reconciliation. We know history is not rational, but we want a better world, [with] rich and successful European nations' (Ibid). Romanian living standards have improved since 1989 and were about two-thirds of those in Hungary in 1996 while unemployment in the border zone is 10.3 per cent in Romania and 8.5 per cent in Hungary (Kukorelli Sz., et al., 2000). Considerable shopping tourism from the Romanian side of the border into Hungary has developed (Ibid). Joint membership of NATO has encouraged military co-operation along the border.

Differences between East and West Border Regions

Table 2.1 provides a statistical comparison of the two study areas. The first section looks at changes over time for the two border counties of Győr-Moson-Sopron in the west and Békés in the east. It is clear that the western county was more prosperous even before the political transition and the gap between the two areas has steadily increased especially in terms of regional income, per capita income and tax base, Gross Domestic Product and unemployment. In terms of enterprises per 10,000 inhabitants, under socialism, when such institutions were almost all government owned, the east had more than the west but by 1997 the west had overtaken the east, although the number in both areas had increased considerably. As early as 1991 the western border area led the country in foreign direct investment per capita and by 1998 this leadership had become entrenched with Győr being at the same level as Budapest while Békés County had the lowest level in the southeastern border region.

In terms of location and quality of life factors influencing economic activity based on composite variables for 1997, regional differences are measured on a four-point scale – high, favourable, moderate and low (Rechnitzer, 2000, pp. 49–50). Accessibility was measured by car and telephone density and the proximity of Budapest and Vienna and showed the dominance of the M1 (the Vienna-Budapest motorway) corridor in attracting foreign capital. Győr-Moson-Sopron County was at the high end of the scale while Békés County in the east was at the lowest point in the scale. Community infrastructure levels were measured by density of homes, proportion of homes connected to a sewage system, electricity consumption and the proportion of households with waste collection. In general most parts of the country were well served following post-transition efforts to improve rural infrastructure although the west was classified as high while the east was only moderate. In the border zone of Győr-Moson-Sopron County income levels were higher, there was more housing construction and more cars per capita than in the county as a whole (Kukorelli Sz. et al., 2000). Average personal tax levels were much lower in the villages of the eastern border than in the western border villages in 2002 further reinforcing the pattern of lower incomes and poorer quality of life in the east than in the west.

Population activity was based on population change 1990–94, migration differences 1980–89 and the number of retired. This factor showed little regional variation although Békés was moderate while Győr was favourable. However, Rechnitzer (2000) asserts that the demographic parameters are deteriorating with an average annual decrease in population of 25 000 to 30 000 which is especially noticeable in the poorer, more

Table 2.1 **Economic and social indicators for east and west border regions of Hungary**

	West	East
County level		
Temporal comparisons		
Regional GDP 1994/inhabitant*	95–104%	80–94%
Regional GDP 1997/inhabitant*	105–117%	65–79%
Unemployment rate 1992	8.0–10.9%	15.0–17.9%
Unemployment rate 1999	3.7–7.0%	11.0–14.9%
Branch banks 1995	47	25
Branch banks 1997	40	19
Foreign direct investment forints/capita 1991**	100–249%	Under 99%
Foreign direct investment in forints/capita 1998**	1000+%	250–499%
Enterprises /10,000 inhabitants 1988***	48	57
Enterprises/10,000 inhabitants 1997	726	564
Personal income tax base/capita 1990*	90–99.9%	80–89.9%
Personal income tax base/capita 1998*	100–116%	70–79.9%
Total income per capita, 1987*	100.1	98.4
Total income per capita, 1995*	100.2	92.8
Comparison of composite variables for 1997		
Traffic and communication	high	low
Community infrastructure	high	moderate
Population activity	favourable	moderate
Education and wage level	high	favourable
Social infrastructure	favourable	moderate
Activity of local economy	high	favourable
Quality of life	high	moderate
Local business assistance	high	low
Border regions, 1998		
Number of individual farmers /1000 inhabitants	74.0	258.0
Number of commercial lodgings/1000 inhabitants	26.0	20.4
Number of guestnights/1000 inhabitants	1,294.1	883.2
Number of cars per 1000 inhabitants	218.7	230.0
Ratio of export income to total net income %	54.1	13.1

* % of national average = 100.
** % of national average in 1991.
*** In 1988 includes industrial and agricultural companies, cooperatives and stores.

Source: Adapted from Figures 1, 2, 3, 4, 7, 11, 12 and Tables 2, 4, 6, and 7 in Rechnitzer, János (2000) *The Features of the Transition of Hungary's Regional System*, Discussion paper No. 32, Pécs: Centre for Regional Studies of Hungarian Academy of Sciences.

marginal parts of the country as young people of reproductive age migrate to the western more prosperous areas. Pál and Nagy (2003) also see this as a problem in border villages with the concomitant aging of the village population. The most recent population figures for the change between 1996 and 2003 also show a decline in the total population of our sample villages and a slight decline in their proportion of the county population, from 3.79 to 3.74 per cent in Békés and from 6.69 to 6.48 per cent in Győr-Moson-Sopron (Statistical Yearbooks for Békés and Győr-Moson-Sopron counties for 1997 and 2004). However, despite an overall national population decline of 0.57 per cent, from 10 174 442 in 1996 to 10 116 742 in 2003, Győr-Moson-Sopron County grew by 3.45 per cent while the population of Békés County fell by 1.9 per cent (Ibid). The relatively stable figures for the proportion of the county population in border villages indicate that, despite the differences between the two counties, the border villages are closely following their county patterns of population change although not growing as fast in the west and declining faster in the east.

The quality of the labour force available in the various counties was measured by proportion of students in higher education, population with academic degrees and the level of wages. Győr-Moson-Sopron was high on this scale and Békés was favourable. In general regional differences in wage levels were marked only in terms of differences between Budapest and the rest of the country. However, Rechnitzer (2000, p. 48) asserts that among hard to measure workforce characteristics such as experience and labour culture, historical regional differences are still found. The index of social infrastructure is based on the numbers of doctors, hospitals and primary schools. The west is not as high on this index as the counties of Baranya and Csongrád in the south but it does display a uniform favourable level while the east is more variable with Békés recording a moderate level.

The local economy's activity level was based on trade turnover, household expenses, prices of accommodation, level of personal income tax and the number of new homes built. The western part of the country has a higher than average value while the rest of the country is more varied although Békés has a favourable rating. Assistance provided to business by local governments as measured by the number of duty free zones, the proportion of settlements with industrial taxes and the amount of support from business assistance centres and state funds is very diverse. Győr-Moson-Sopron County was the only area apart from Budapest with a high ranking on this index. Békés County, on the other hand, was at the bottom of the scale. Finally a subjective locational factor of quality of life was measured based on the number of protected natural areas, the number of theatre and exhibition visitors, security as measured by number of crimes and the number of civil society organizations. Győr-Moson-Sopron County ranked high on this scale with a rich cultural milieu, major protected areas and strong civic traditions. Békés County, however, only ranked as moderate on this scale. According to Rechnitzer (2000, p. 51) in the west, especially in Győr-Moson-Sopron, 'the quality of life, as an indirect influence factor to location, are [sic] unequal in the country'.

Comparative economic activity statistics for the western and eastern border regions in 1998 show distinct differences (Table 2.1). The eastern region has a much higher proportion of farmers and cars than the west suggesting that it is still a largely agricultural and isolated region dependent on private cars for transportation between settlements. The western border region has one third more tourist bed-nights than the east and more tourist accommodation and it obtains a much larger proportion of its regional net income from exports than occurs in the east. These differences show that the Austrian-Hungarian border region has an outstanding level of economic activity as a consequence of multilateral co-operation. In the south-east Romanian border region, although there is some industrial production, export potential is low because both sides of the border have few resources and are peripheral to their respective countries. These differences are reflected in the number of border crossings in the two case study areas.

Border Crossings

Border crossings can vary according to the types of transportation used, the hours and seasons of opening and the restrictions on people of different nationalities (Figure 2.1). Open crossings in our study areas were very few in 1989 but have been increasing rapidly both in number and in the variety of traffic allowed (see Figure 2.1). Many of the early crossing points have been closed intermittently.

The crossing at Hegyeshalom for non-motorized traffic was opened in mid-1950. It was motorized until a new wide road was built which became a motorway and then the old crossing point was closed to avoid environmental pollution in the border villages. The road crossing at Gyula, in eastern Hungary, was opened in 1970. The crossing at Hegyeshalom was expanded in 1982 and one at Kópháza for passengers and cargo opened in 1985. Most crossings in the west were opened since the fall of the Iron Curtain. The temporary crossing between Dombegyház in Hungary and Variaşu Mic in Romania had been planned as a major crossing on the proposed Budapest-Bucharest highway but construction of this route has been postponed (Lelea, 1999). The counties of Arad and Békés have agreed to open this crossing for trade in agricultural goods only but national agreement is awaited (Kiss, 1999). Three border crossings in the study areas date back to the 1920s and two of these are major rail crossings in Sopron and Lőkösháza (Figure 2.1). Two crossings were opened in 1950, one in 1970 at Gyula in the east, and two in the 1980s. Despite the presence of the Iron Curtain in the west, four of the five crossings opened after the Second World War and before 1989 are on the western border. Of the current ten border crossing points in the west in Győr-Moson-Sopron County, nine have been opened or expanded since 1990 (Figure 2.1). The most recent ones, opened in 2001, are the road crossing at Zsira and the water crossing at Fertőrákos open only from May to mid-October. A border crossing for non-motorized traffic open only between April and November was also opened at Fertőrákos in 1992. The water crossing is only available to Hungarians and citizens of neighbouring countries while the road passenger crossing may be used by people of all nationalities for whom visas are not required. Clearly both these crossings are mainly for the use of tourists, especially the birdwatchers who flock to the lake and its surrounding marshes in the National Park, as this area is a major flyway for migrant birds. At the other extreme of uses is the container port at Hegyeshalom opened in 1994. In the east six of the eight border crossings have been opened since 1991 but three, Körösnagyharsány, Elek and Dombegyház, are temporary, only opening on a few holidays each year, as requested locally. On the Austro-Hungarian border the Pan-European Picnic crossing at Sopronpuszta opened in September 1989 and the crossing at Albertkázmérpuszta opened in 1997 near

Jánossomorja, are also used on only a few special occasions each year. The wooden bridge at Andau, near Jánossomorja was rebuilt after the transition. It is a famous place, because during the revolution in 1956, many people fled from Hungary across this bridge. This bridge, re-opened in 1999, is located in the territory of the national park, so it is now a crossing point for visitors and is available only during daylight hours in summer.

Road crossing points vary from the major crossings on highways used by trucks and other road transport to crossings which are only accessible by bicycle or on foot. The four non-motorized pedestrian-only crossings in the west have been opened at various times between 1950 and December 2001. In western Hungary we were told by the mayor of one village in 1997 that the neighbouring Austrian village would not allow the crossing between the two villages to be open to cars as they did not want to have old, polluting Trabants coming into Austria. A similar situation exists currently at Hegyeshalom where the authorities in the Austrian village of Nickelsdorf do not want to open the border to motor traffic because of pollution. Local people want this crossing to be motorized as the nearby main border station on the motorway becomes very busy and has many delays, especially at holiday times. However, commuters have overcome this restriction by owning cars on both sides of the border. At Zsira, there is an Austrian spa and hotel right at the border. Many of the employees at the spa come from the Hungarian side of the border but can only cross on foot or on bicycles. This crossing is closed at night with the hours of closure varying with the seasons.

Many border villages had traditionally celebrated joint feast days. Some of the old footpaths between the villages are now being opened once or twice a year so villagers can meet once again for these joint celebrations. In other cases, the border has cut through landholdings and prevented farmers being able to access part of their land, as we were told by an elderly farmer in eastern Hungary. Since he had owned quite a large area of land and had been imprisoned as a *kulák*, he had little recourse. The Trianon border in eastern Hungary cut eight railway crossings on the Romania-Hungary border (Kukorelli Sz. et al., 2000) and these railways still 'dead-end' at the border (Plate 2.2).

With the accession of Hungary to the European Union, border posts have to meet EU regulations. In the last couple of years many of the major crossing points have upgraded their facilities with large parking areas, restaurants and souvenir shops. On the western border Western Europeans and Hungarians have been able to cross very easily by just showing their passports for several years. Access to Romania, however, was limited by the need for visas for Western Europeans until quite recently. The sale of visas at the border provided an opportunity for border guards to make black market profits. Such illegal profiteering made jobs as border guards much sought after in Romania.

Attitudes to the Border Location

Not surprisingly, the more open border in the west encouraged flows of tourists and trade into western Hungary. The border in both east and west has also become the site of increasing criminal activity. Recently prostitution has become less visible along

Plate 2.2   Railway line between Hungary and Romania blocked after the Trianon Treaty

the roads near the western border but also occurs around the main border crossings into Romania in south-eastern Hungary. Migrants from poorer countries to the east, such as Ukraine, still come to Hungary to service truck drivers and others who frequent main border crossing areas. The pimps who control the prostitutes are often also involved in drug trafficking although this takes place at the more minor crossing points. The other main illegal activity is that of smuggling immigrants into the European Union. On the Hungarian-Romanian border, the presence of illegal traders and their catering, accommodation and health problems cause difficulties for border settlements (Pál and Nagy, 2003).

The growth of criminal activity on the border is considered as the major disadvantage of living near the border. Quiet, isolated villages find themselves dealing with crime on a daily basis. In eastern Hungary most of the problems we were told about involved petty theft for which illegal Romanian immigrants were blamed. In the west, on the border with the European Union, criminal activity was more serious and often involved firearms. After 1 May 2004, when the frontier with the European Union lies along the border between Hungary and Romania, it is expected that crime will increase in this area while decreasing in the west.

In the Győr-Moson-Sopron County area it was predicted that cross-border trade and flows of capital and tourists would encourage the growth of entrepreneurship while in Békés County the presence of the border with Romania was expected to have a neutral effect on the growth of small firms. We investigated the attitudes of entrepreneurs to their border location.

The northwest has fewer foreign residents than the southeast but both had net outmigration in the mid-1990s (Geographical Research Institute, 1994 and 1995). Of the border villages in our sample, three out of ten had a population increase between 1996 and 2003, while in the west nine out of 17 grew. However, by the late 1990s we were meeting individuals who were choosing to move to the western border villages from central and north-eastern Hungary because they were aware of the economic benefits available. This area had the second highest stock of foreign capital investments after the capital, Budapest, because of the following regional characteristics:

- excellent cross-border transport connections,
- strong work ethic based on past experience of industrial production,
- pre-existing diversified international production and sales connections,
- good infrastructure and pleasant living environment,
- a high level of local support for adjustment strategies in terms of preparedness for EU accession, availability of green field investment opportunities, establishment of industrial parks, and a favourable local tax policy (Rechnitzer, 2000).

For individuals, the western border offered easy commuting to well-paid jobs in Austria, as well as proximity to western clients and tourists. In our 1998 survey of entrepreneurs six men and six women told us that they saw the western border as an opportunity. Western visitors became the main clientele of many service businesses ranging from car detailing and repairs, to dentistry and cosmetology. They were also the basis of the fast growing tourist industry in the western border area focused on the bi-national Park and the baroque Eszterházy Palace where Haydn spent many years. This palace and its surrounding landscaped gardens were being restored in 2003 and chamber music concerts were already being held there.

In eastern Hungary, villages felt isolated as the border had cut them off for many years from their traditional urban spheres of influence in Romania. Even with a more open border, there was little interest in cross-border cooperation, except at the individual family level, as Romania was not seen as a market. Only two people in our survey saw a location on the eastern border as a source of opportunity and one of them was a Romanian immigrant. The nearest urban centre for most villages was Békéscsaba and there was a general feeling that the roads linking them to this centre were inadequate and public transport had declined in availability and increased in cost since the transition. This feeling of isolation was well expressed by an architect in one village who told us that he felt that he lived 'ten kilometres beyond the end of the world'.

Many villagers were trying to find ways of moving to one of the towns in Békés County. They felt that this was especially important for their children so that they could have better educational opportunities. This has led to many rural dwellers investing in flats in the towns, ostensibly for children, but often rural women indicated that they too would prefer an urban environment.

In Békés County there has been a steady population decline since 1960 and the proportion of the population living in rural areas fell from 77 per cent in 1960 to 58 per cent in 1980 but has stabilized in this decade, being 39 per cent in 1990 and 38.8 per cent in 1996 (Timár and Velkey, 1998). As capitalism produces differential incomes and land prices, urbanites in eastern Hungary are building larger homes in the villages near to major towns such as Békéscsaba and commuting, while the rural poor are staying there because housing is cheaper and living costs lower. Suburbanization has had little impact on the more remote villages. In western Hungary, there are signs that counterurbanization is taking place as the middle-class retire to their former rural second homes and working age people move to the countryside in search of cheaper housing and more space for families. In the west foreigners are also buying second homes in remote villages which are attractive because of their peacefulness and visual beauty. They are becoming contested sites of the rural idyll. Border villages with easy access to Vienna are especially popular and old cottages are being renovated and modernized. The eastern border does not attract many people in search of second homes because of its isolation. However, a few rich Hungarians from the Budapest area, who were able to acquire large landholdings after the transition, have moved there in search of wilderness and access to hunting. This unspoilt environment and availability of deer and wild boar, also attracts a few foreign hunting parties. Thus conditions for the development of rural enterprises are very different in the two regions but outmigration may be almost balanced by immigration in both regions leading to little recent population change in the study areas.

The Importance of the Border to Local Residents

In Békés County in eastern Hungary, villagers generally felt isolated and cut off by the border as they saw few opportunities in Romania. However, a few were employed as border guards, customs officials or in passport control, or were training for such positions. It is often suggested that these positions are fairly profitable because of the opportunities for bribery and corruption offered by trans-border trade. Such opportunities have probably decreased considerably with the advent of European Union regulations at borders. In fact two of our interviewees stated that they had given up jobs in border control in the mid 1990s because they did not like it and had then become entrepreneurs.

In western Hungary none of our interviewees mentioned working in border control but seven worked as customs agents, probably reflecting the greater flow of trade across the border with Austria.

In the village of Lökösháza which is a rail centre for international trains from Budapest to Arad and where trans-border buses also stop, at least one woman entrepreneur had recognized the opportunity her border location provided. She took over a former gypsy pub and planned to open a food shop, restaurant and guesthouse when the border road crossing was expected to open in 2002. There is still no road crossing at this point (see Figure 2.1). One other person had opened a snack bar at the railway station hoping to serve the Polish, Chinese and Romanian traders who came through. She did not want to expand her business because of the constraints imposed by health and safety regulations. So of the nineteen entrepreneurs interviewed in Lökösháza only two women stated that they recognized the importance of the border as a source of customers and only one of these two women, the one aged 55 in 1998 and not the one of only 29 years, was looking to the future and wanted to expand her business.

On the border with Austria it has been suggested that there is a complete lack of interest in the population on the other side of the border.

> People on either side incorporate the border as a physical barrier in a discourse of past and present limitations to their lives. It provides 'a foil for expressing daily hardships and limited economic means on the Hungarian side, and for articulating anxiety about violence, immigration and the stigma of cultural "backwardness" in Austria'.
>
> (Meinhof, 2002, p. 13).

In their chapter on two villages on either side of the Austro-Hungarian border, south of our study area, Doris Wastl-Walter et al. (2002) show that there is a total lack of communication between the settlements. The mayors of both villages had planned to open a border crossing after 1989. This would have been welcomed by the Hungarians but in a local plebiscite the people of the village on the Austrian side of the border managed to block the plan. 'The eight decades following the establishment of the Trianon borders proved sufficient for the collective memory to become vague about the shared past and for the radical loss of interrelated narratives' (Wastl-Walter et al., 2002, p. 92). They also report a comment by one of their informants that the border had more disadvantages than advantages and 'We were far from everything' (Wastl-Walter et al., 2002, p. 80). This is very similar to comments by some of our informants on the eastern border of Hungary.

The people in our survey, although asked specifically about the advantages and disadvantages of the border, made few comments. In the west five men and three women thought it was an advantage while in the east two men and one woman saw it as advantageous. In the west some people had moved to the border region specifically to set up their businesses because they saw it as an area of opportunity. One man and one woman in the east complained about the border region as a problem area but most entrepreneurs were fairly neutral in their attitudes towards their border location.

Although some people in Austria, especially in south Burgenland, did not want to create new border stations as they felt threatened by cheap Hungarian labour, in our study area relatively close to Vienna, cross border interaction has increased very quickly in terms of tourists and shoppers into Hungary and workers commuting into Austria. The closed nature of the western border during the socialist years protected the environment so that its unspoilt nature is very attractive to tourists today. This is seen particularly at such sites as the bi-national National Park and the Eszterházy Palace and also in terms of quiet bicycle paths between villages, in which it is easy to find food and accommodation in modern establishments (Plate 5.1).

Borders as Sites of Transnationalism

Transnationalism in its broadest form is defined as 'multiple ties and interactions linking people or institutions across the borders of nation-states' (Vertovec, 1999, p. 447). In addition to its use as a theoretical construct for understanding migration, transnationalism is also useful in examining ethnic diasporas (Brah, 1996), capital flows, economic networks and transnational corporations (Beaverstock and Boardwell, 2000; Hudson, 2001; Guarnizo, 2003), transnational commodity culture (Brettell, 2003; Crang, Dwyer and Jackson, 2003; Hitchcock, 2003), transnational livelihoods (Bebbington, 2001; Bebbington and Batterbury, 2001), transnational communities (Portes, 1996; Conway and Cohen, 1998; Voigt-Graf, 2004), transnational space (Voigt-Graf, 2004), and identities (Mitchell, 2002; Nyberg Sørensen, 1998; Kong, 1999; Foster and Froman, 2002) and tourism as transnationalism (Torres and Momsen, 2005). Thus the concept can be seen as being relevant to an understanding of trans-border relationships in many different ways. In relation to the growth of entrepreneurship, transnational commodity culture, capital flows, economic networks, and livelihoods are the key to the success of borders as locations for small business activities. The concepts of transnational ethnic diasporas, space and identities are very important in supporting the growth of social capital in border areas.

A hundred years ago Hungary was described as if 'with its sharply defined natural border, it has been destined by Nature herself to be a uniform state' (de Vargha, 1910, p. 1). 'The orological and hydrographical conditions impart a peculiarly uniform character to the territory of Hungary' (Ibid). It was seen as unchanging, 'the Hungarian Kingdom is one of the oldest states in Europe. With its present area, practically with the boundaries of its frontiers of today, it has existed, as a united National State, for over a thousand years' (de Vargha, 1910, p. 7). Yet despite this apparent geographical uniformity, in 1900 barely half the population were Magyars by tongue while another 8 per cent non-Magyars spoke Hungarian, 16.6 per cent had Romanian as their mother tongue, 11.3 per cent German and the same percentage were Slovak (de Vargha, 1910, pp. 17–18). In the smaller Hungary of the twenty-first century almost all the population speaks Hungarian but other languages are also still spoken, especially in border villages. The boundary changes brought about by the Treaty of Versailles and after the Second World War resulted in many communities that had once been in Hungary finding themselves on the other side of the border. The western border area today has German and Serbo-Croat minorities while in the East the main minority group is Romanian with a few Serbian-speaking people in the southern part. A passive form of transnationalism was created by redrawing political boundaries. Only in the post-socialist era has an active transnationalism been able to develop.

This active transnationalism takes the form of trading, working and visiting. The joint history of Hungary's trans-border areas has encouraged contemporary tourism in western Hungary and western Romania, encouraged by language commonalities, creating what Appadurai (1996) recognizes as special 'translocalities'. 'Many such locations create complex conditions for the production and reproduction of locality, in which ties of marriage, work, business and leisure weave together various circulating populations with kinds of locales to create neighbourhoods that belong in one sense to particular nation-states, but that are from another point of view what we might call *translocalities*' (Appadurai 1996, p. 192). The growth of these border translocalities is encouraged by cross-border agreements for common planning and development, such as the PHARE-CBC and the INTERREG programmes.

Conclusion

The border regions played a major role in the Hungarian post-socialist transition. Development occurred most successfully in those border regions with easy accessibility to the economic centres of Western Europe leading to growing trans-border connections in terms of partnerships and membership in European organizations. Thus the transition 'rearranged the spatial structure of Hungary' and the gap between the west and east of the country increased (Rechnitzer, 2000, p. 60).

Chapter 3

The Transition in Rural Areas

The positive impact of the transition reached rural areas later than urban areas and varied according to the alternative opportunities that became available in different parts of the country. As urban firms downsized, the first elements to be eliminated were workplace services such as childcare facilities and bus services for workers commuting from rural areas. In rural areas wages are generally lower than in urban areas and jobs became especially hard to find after the agrarian reform (Répássy and Symes, 1993). The impact of rural restructuring as the large village-based cooperatives and state farms collapsed, was far reaching.

Standing (1991) notes that in the Soviet Union the end of collectivization in agriculture and the transfer of labour from agriculture to the tertiary sector put considerable strain on the labour market. Moghadam (1992) suggests that this rural restructuring has profound gender implications and draws parallels with China where the retreat from collective farming to the household responsibility system and the family farm, encouraged a resurgence of patriarchal attitudes in the countryside and a diminished status for women. These changes often forced women to give up their jobs in order to look after their children at home. Private employers could not afford to offer the subsidized childcare services formerly provided by the state which made female workers more expensive than male (Gömöri, 1980). Employment in agriculture fell rapidly from 17.4 per cent of the labour force in 1989 to 9.3 per cent in 1993 and has continued to decline from 7 per cent at the end of 1999 to 6 per cent in 2002 (Kiss, 2003; Economist, 2002 and 2005).

Before the Second World War, Hungary was characterized by rural overpopulation with more than half the nation's people dependent on agriculture. In 1944, after what Dumont (1957, p. 466) described as 'the most half-hearted measures of agrarian reform to be found anywhere in eastern Europe between the wars', Hungary was still dominated by very large estates to a greater extent than any other country in Europe. Of these huge estates, three hundred averaged 6883 hectares and the Eszterházy family, which had possessed some 425 100 hectares in 1914, still owned 119 430 hectares in western Hungary and 58 700 hectares just across the border in Austria. At the same time, three-quarters of the country's farms had less than three hectares, accounting for only one-fifth of the country's total area (Dumont, 1957). Such stark contrasts in land ownership and the new political direction led to the setting up of cooperative farms and the implementation of the Soviet collectivization system. By the end of 1948 there were 350 cooperative farms, with 1500 a year later, and rising to 5315 by the end of 1953. Together with the State farms which made up 12.7 per cent of the cultivated land, the cooperatives occupied 37.3 per cent for a total of 50 per cent for the socialized portion of the agricultural area (Dumont, 1957). The average cooperative at this time had only 240 hectares of cropland but supported about 2.5 workers per hectare, largely because of continuing rural overpopulation. There was resistance to collectivization on the part of many peasants and food shortages were widespread (Dumont, 1957). Those who owned more than twelve hectares of farmland, or threshing machines or shops were considered *kuláks* and were excluded from cooperative formation. They were represented as class enemies. Thus those with the most knowledge about running farms were excluded from the new agriculture and production fell drastically in 1952 (Asztalos Morell, 1999). As can be seen from Table 3.1 the socialist sector dominated agriculture by 1961 although there was a slight decline in the late 1980s. If the household plots of cooperative members are included as private agriculture, then the proportion of family farms declined from 17.2 per cent in 1961 to 13.1 per cent in 1984 and increased to 14.9 per cent in 1989.

Table 3.1   Percentage distribution of agricultural land according to sector for the years 1953, 1957, 1961, 1984 and 1989

Sector	1953	1957	1961	1984	1989
State sector	23.3	22.2	17.9	15.1	14.8
Co-operative sector	22.8	11.4	74.5	77.6	74.8
Socialist Sector	**46.1**	**33.6**	**92.4**	**92.7**	**89.6**
Private Sector	**53.9**	**66.4**	**7.6**	**7.3**	**10.4**

Source: Adapted from Asztalos I. Morell (1999) *Emancipation's Dead-End Roads*, Uppsala, Sweden: University of Uppsala, Studia Sociologica Upsaliensius 46, Table 3.5.

The proportion of the population dependent on agriculture fell just slightly between 1930 and 1949, from 56 per cent to 54 per cent, but by 1960 had declined to only 37 per cent falling rapidly to 15.4 per cent in 1990 and a mere 8.2 per cent in 1996 (Kovács, 1999). In 1987, 10.5 per cent of the total personal income of all Hungarian households and 18.6 per cent of the personal income of village households came from the agricultural second economy. By 1990, 41 per cent of the total agricultural production was produced by the few private farms and nearly 1.5 million household plots (Andorka, 1993). Szelényi (1988) hoped that a new bourgeousie would evolve from this rural second economy but after the transformation of the regime in 1989, agricultural activity rapidly declined, accounting for only 3.8 per cent of GDP in 2004 (Economist, 2005) as compared to 13.7 per cent in 1989, 7.0 per cent in 1995, and 4.1 per cent in 2001 (Kiss, 2003).

The Privatization of Land in the Transition Period

A direct consequence of the change in the political system was the establishment of the market economy which demanded a new form of land ownership structure. This was one of the goals of the first post-1989 elected government, especially in relation to agricultural land. The populist Independent Smallholders Party was a member of the governing coalition and supported the privatization of land, although the main governing party did not. Nevertheless, the issue of compensation and privatization of arable lands became politically very important. It was decided that privatization should be achieved through liquidation of the large socialist holdings and the creation of private farms or of new types of cooperatives. This change was carried out with three Acts of Parliament in the early 1990s: the First Act on Compensation, The Act for the Distribution of Agricultural Lands and the Act of Transformation. This was the third restitution of land in Hungary in the twentieth century.

The break-up of state agricultural lands was carried out through the issue and sale of compensation vouchers. The First Act on Compensation, Act No. XXV of 1991, laid down the principle that the property losses caused by collectivization should be compensated for in the form of vouchers, in a regressive way with five million Hungarian forints set as the upper limit of compensation. Compensation vouchers functioned as securities which could be used for the purchase of land through competitive bidding or land auctions. Before the transition, the agricultural cooperatives had properties of three different types:

- Lands owned by members of the cooperative;
- State-owned lands that the cooperatives used free of charge for an unlimited period of time;
- Lands owned by the cooperative under Act No. IV of 1967.

The process of compensation concerned all three forms of property. The different types of property were linked to different land funds: the members (employees) land fund, the compensation land fund, and the land fund that included lands owned by the state but used by the cooperative.

The compensation of land started in 1992 and was mostly completed by 1995, although there are still claims in the courts by some who remain unsatisfied. Those who held compensation vouchers were entitled to participate in so-called competitive bids for land. Land reallocation committees were organized in each cooperative responsible for the designation of the pieces of land belonging to members of the cooperative and those pieces available for compensation.

It became clear, that if, according to the 1991 Act, the winners of the restitution of land lost in 1947 and the losers in the establishment of cooperatives in the 1950s and early 1960s, were to be compensated then approximately two-thirds of the new land owners would be urban residents with little involvement in agricultural production. This meant that land owners and land users would form two separate groups. It was also feared that the fragmentation of land ownership would reduce the competitiveness of Hungarian agriculture. Both predictions proved to be true. According to Karalyos (1997, 72)

> ... in the changing ownership of arable land, the economic aspects of privatization – that favour the establishment of more effective forms of land cultivation and more effective business forms – did not play a role, it was basically political considerations and objectives that dominated in the regrouping of the resources to the detriment of the co-operatives and in favour of the social groups eligible for compensation and able to assert their interest in the transition period.

The factors leading to the disintegration of the cooperatives were compensation on the one hand, and the distribution of the land and property among the individual members on the other. According to Act No. II of 1992 on the foundation of cooperatives, the property of the former cooperatives had to be distributed: 40 per cent of the property was to be distributed among the active members of the cooperative; another 40 per cent was given to pensioners of the cooperatives and the remaining 20 per cent to external owners. The distribution of the property allowed the members to leave the cooperatives with their own designated land until the end of 1992. Those who seceded from the cooperative could either cultivate their land on their own as private farmers or establish other cooperatives. These measures reinforced the weakening and disintegration of the former agricultural cooperatives but no mass disintegration of the cooperatives took place. Hungarian cooperative members realized that they had a lot to lose by leaving the collective and the initial response by most farmers was to stay with the cooperative. After the Act came into force, 7000 cooperative members and employees decided to secede from the cooperative collectively and almost 9000 left individually. In total only 5 per cent of members left the cooperatives taking 1.5 per cent of the land used (Laczó, 1994). Group secession was unusual but occurred in at least one area under the influence of the local leader of the Smallholders Party (Swain, 2001).

Surveys revealed that the seceding cooperative members were often already cultivating their own land and in many cases they had leased land. The Act for the Distribution of Agricultural Lands (Act No. II, 1992), designed by its creators to serve the fundamental transformation of agriculture, favoured the existing management of cooperatives. They saw this Act as a tool for the restructuring of cooperatives and, using their formal and informal relationships, they became the winners in this transformation of agriculture. In most cases the former managers of the cooperatives became the leaders of the transformed co-operatives. The 1992 Act induced the distribution of the property of 1300 state cooperatives, worth a total of 26 billion Hungarian forints (Csite and Kovách, 1995). One-third of the property was taken over by the active workers employed on the cooperative, and 30 to 40 per cent by pensioners of the cooperative. The remainder was given to outsiders and so cooperative land ownership ceased to exist. Nationally, by 1994, 31.7 per cent of land was farmed by cooperatives, 35.9 per cent by private corporate farms and 32.4 per cent by individual farmers (Swain, 2001).

After this transformation, the property of the former cooperatives was operated by the active members whose interests were not the same as those of other land owners. While the active members were interested in the re-investment of profits in the land, the outsiders sought businesses and investments with higher profits. The cooperatives, having lost half of their former cultivated lands, were forced to lease land, and the costs of leasing decreased the amount of capital available for mechanization and other investments to improve efficiency. Medium, or in many cases short-term leasing of land, does not favour intensive cultivation as the return on input costs may be over a longer period than that of the lease. It is also unsuited to stock rearing because of shortage of land for fodder.

The agricultural cooperatives were weakened in the first place not by the First Act of 1992, nor by the number of seceding members as a consequence of the Act on the Distribution of Agricultural Lands, but rather by the worsening conditions of their operation, exacerbated by the need to lease lands, declining state support and the loss of markets in the former USSR. The new bankruptcy laws, introduced by the Cooperative Transformation Act of 1992, led to 28.8 per cent of those existing in 1989 declaring themselves bankrupt and having to be liquidated (Swain, 2001). The cooperative transformation process became one of co-operative breakup (Ibid).

However, this large scale liquidation did not decrease the number of cooperatives. In fact the total number increased with the transformation and re-establishment of the former cooperatives, as several cooperatives, each involving one village, replaced the former mammoth multi-village cooperatives, or the latter were transformed into limited companies. These new forms of cooperatives did not fulfil the same role in the rural areas as the old cooperatives had done before 1990 particularly in terms of employment, and so the restructuring of the cooperatives radically changed village life.

State farms were also privatized. In 1991, 124 state farms were in operation, occupying 976 000 hectares, 36.6 per cent of state-owned land. The privatization of the state farms provided an opportunity for improvement in the efficiency of their operations. The total land area of the state farms gradually decreased after the start of the privatization process, because of land distributed among their employees worth 20 golden crowns per person. The total area in state farms fell to 100 000 hectares after privatization. The state farms faced solvency problems as a consequence of the difficulties of paying back the credits taken out in the 1980s and because of their accumulated losses. In 1989 some 13 state farms showed a deficit and 45 were in this financial situation by 1991.

The Act of Transformation of 1992 allowed the state farms to be transformed into limited liability companies or shareholding companies. The latter was the most popular form adopted. The transformed state farms did not own land as they had been deprived of their land holdings by the State Property Agency. This Agency became the new owner of the lands and the state farm companies had to lease land from it. Although this process was different from the privatization of the agricultural cooperatives, in the period of decentralized privatization the compensation vouchers became more important in the case of state farms as well (Csete, 1995). As a result of the privatization process, of the 947 000 hectares of state-owned land in 1990, only 417 000 hectares were still state-owned in 1997. By 1995, some 46 state farms had been privatized. Nine state farms were privatized in 2004 but the process is not yet completed. According to the Hungarian Privatization and State Holding company in March 1998, approximately eight million hectares of arable land in Hungary had been privatized (Lovászi, 1999).

Privatization of land is now almost complete, and there were almost two million land owners in Hungary, with an average farm size of three hectares in 1999. Eleven per cent of all land owners have holdings of less than one hectare and 60 per cent have farms of less than ten hectares. In addition to the changed ownership structure, farm structure has also changed. The number of private farms in 2000 was 958 534 while there were also 8382 agricultural companies. The average size of the private farms is 4.15 hectares, while the average size of the lands cultivated by companies and cooperatives is more than a hundred times that at 452.3 hectares. These figures demonstrate the divided nature of the agricultural sector in Hungary, together with the vast differences in the conditions of farming between the two sectors which affect employment possibilities in villages, rural income levels and the future of rural spaces.

Land Privatization in the Study Areas

There was marked regional differentiation in responses to the privatization of land in Hungary. The quality of arable land in both Békés County and in Győr-Moson-Sopron is above average and in the former area is amongst the best in the country. In both study areas, the land is capable of high agricultural productivity and can be competitive by European Union standards (Kovács, 1999). Half a million people bought land at the privatization auctions with demand being highest in the counties of the south-east, including Békés. Kovács (1999) suggests that the demand for land in Békés was high because this region had a more highly developed private agricultural sector as early as the 1970s and is dominated by agricultural towns, large villages and scattered farms with a long tradition of horticulture. This county also supported the Smallholders Party which was the party which first encouraged the restitution of lands to their original owners, and is a largely underdeveloped agricultural region. In Győr-Moson-Sopron there was also considerable interest in reclaiming land as this was the first county to have cooperatives established and so there was still bitterness among those who had lost their land at the time of collectivization. Interest in land in this area was also probably reflecting the pattern noted by Kovács (1999) of high interest in land purchasing in areas near towns and vacation areas. She notes that citizens of large towns are better informed than the rural populace and so bought land as an investment in those areas where they expected values to rise fastest following EU accession. In addition, Kovács (1999) notes that it was probably demand from foreigners for land, usually bought through 'pocket contracts' by buying compensation vouchers from local people very cheaply, which made locals value it more as a market commodity, especially along the border with Austria.

So both our study areas had a high proportion of their population involved in the privatization of land, reaching 40.4 per cent in Békés and 31.7 per cent in Győr-Moson-Sopron. There has been a deconcentration of agricultural land but this has now been reversed as many elderly or unskilled people sell or rent their land back to the surviving cooperatives or to farmers looking to expand their private holdings. The majority of the new owners do not actually cultivate their land but Kovács (1999) concludes that it is the new owners and those already farming who have significant peasant roots and a good education who will be the main representatives of the future agricultural bourgeoisie.

The Transition Across the Border in Romania

In 1989, Romania had 27 per cent of its population involved in agriculture compared to 20 per cent in Hungary (Lelea, 2000). After the 1989 Revolution the process of modernization of agriculture stopped and began to regress, although the decline in the use of agricultural chemicals was beneficial to the environment. The cooperatives were broken up in 1991 and the land returned to its former owners. The increased fragmentation of land made mechanization more difficult, leading to an increasing substitution of labour for machinery. Consequently, by 1997, unlike the situation in Hungary, the population involved in agriculture had increased to 35 per cent (Lelea, 2000). Some of the new land owners did not want to farm and they were allowed to donate their land to an 'association', a new type of cooperative, and in return became members and were reimbursed with a share of the crops produced, or occasionally with cash. This arrangement also was found in Hungary for those who rented back the land they had got from the cooperative after the land reform. As in Hungary, those who had been part of the management of the former state farms gained most from privatization. In a village in western Romania, a woman who had been an agricultural engineer in the cooperative operated 35 hectares of land and owned her own tractor (Lelea, 2000). She was the only woman agricultural entrepreneur in the Romanian border zone near our study area in eastern Hungary.

Although the soils in western Romania are as good as those across the border in Hungary, Romanian agriculture is very inefficient and farmers complain that they cannot compete with cheaper imported Hungarian wheat. Most of the new small private farmers and those who receive crops from an association use the farm products for family subsistence. There are few non-farm jobs available but this is gradually changing. In the town of Curtici, near the Hungarian border, there is a very efficient vertically integrated 'association', which has been helped to produce such value-added goods as yoghourt, by Italian capital and training (Lelea, 2000).

Gender Divisions of Labour in Agriculture

During the period from the reorganization of cooperatives up to the 1970s, collectivization was to a degree flexible which allowed for self-selection of leaders at the village level (Asztalos Morell, 1999). This enabled the transfer of some of the patriarchal peasant family structures into the cooperatives, preventing women's movement into positions of political and economic leadership. Members of the former landowning peasantry took on the more prestigious jobs such as team leader or cart driver, while those from the former servant class moved into jobs in crop cultivation or animal husbandry (Asztalos Morell, 1999). However, collectivization was also accompanied by deskilling under which the former peasants became wage workers and according to Asztalos Morell (1999) this meant a demasculinization of the status attributes of the former head of household. In 1982 only 1.1 per cent of presidents and vice-presidents of cooperatives were women, a decline from 3 per cent in 1971, as a result of the mergers of cooperatives. Women made up the majority of administrative workers but few were production supervisors (7.2 per cent of women compared to 50.4 per cent of men in 1982). However, the family remained as a production unit with women, as family helpers, working to meet seasonal labour demands while cooperative members worked year round. Table 3.2 shows that as late as 1990 only 31 per cent of cooperative members were women, while women made up 87 per cent of family helpers and 32 per cent of employees. In order to qualify for benefits from the cooperatives a certain number of work days had to be completed so the regularity of employment meant differential access to cooperative rights and benefits as well as differences in wage levels. Overall, the degree of feminization of the workforce was highest in the two 'reserve' labour categories: family helpers and seasonal workers. In contrast the proportion of women was lowest in the 'core' categories of cooperative members and permanent employees.

Men and women on the cooperatives tended to occupy positions similar to those they had held on the old feudal estates. Dealing with horses had been a male occupation

Table 3.2   Distribution of active earners in the agricultural sector according to employment status and percentage of women (1960–1990)

Employment status	Distribution of workers				Proportion of women			
	1960	1970	1980	1990	1960	1970	1980	1990
Cooperative member	28.7	58.7	49.2	42.1	29.1	32.9	30.3	30.8
Family helper of coop member	3.0	8.1	6.4	0.3	57.9	93.9	97.7	86.6
Employee	16.8	26.6	38.0	50.5	20.4	28.8	28.5	31.7
Socialist sector combined	*48.5*	*93.3*	*93.6*	*92.9*	*27.9*	*37.1*	*34.2*	*31.4*
Private farmer and family helper	51.5	6.7	6.4	7.1	48.0	58.9	65.1	25.75
Total	**100.0**	**100.0**	**100.0**	**100.0**	**38.2**	**38.5**	**36.1**	**31.2**

Source: Adapted from Asztalos I. Morell (1999) *Emancipation's Dead-End Roads?* Uppsala, Sweden: University of Uppsala, Studia Sociologica Upsaliensius 46, Table 10.9.

and this was carried forward to male domination of machinery-using tasks. As late as 1990 only 0.4 per cent of tractor drivers were women. Over half the workers involved in the growing and nurturing of plants were women. Overall, women constituted only 11 per cent of skilled workers but 26 per cent of unskilled workers on the cooperatives (Asztalos Morell, 1999, 403). With modernization and technological change, women workers became concentrated in those agricultural sectors least affected by these changes such as the raising of small animals and growing of ornamental plants, and in non-agricultural activities such as small scale industrial production. In Győr-Moson-Sopron County, where the industrial sector was well developed, many men gave up working in the cooperatives and became commuters to jobs in nearby towns before the transition. In these cases the woman of the family became the member of the cooperative in order to preserve the propriety rights of the family as original owners of the land.

In the early period of collectivization, 1956–1968, household plots were used for subsistence and women provided almost three-quarters of the labour time. With increasing market orientation of household production men became more involved and women less so (Table 3.2). Asztalos Morell (1999) argues that women saw working on the family plot as an extension of their household reproductive duties while for men the increased economic value of such production, especially as it became more important in the black economy, allowed them to reassert their breadwinner role. For the most dynamic producers technical and administrative knowledge gained on the cooperatives was extremely important in laying the foundation for post-transition agricultural businesses.

Time budget studies provide evidence of the important role of women in small plot production showing that men over 15 years of age worked on average 2.5 hours per day on their plot and women two hours despite their heavy responsibility for household chores (Tóth, 1992) although Barta et al. (1984) report that men and women in 1963 and 1976/7 both spent an average of 42 minutes a day on household farming. Among these households operating in the secondary economy there were more commuting workers than peasants (Tóth, 1992).

Although the number of people living in villages fell from 52 per cent in 1949 to 38 per cent in 1990 of the national population and the proportion making a living from agriculture fell even faster, over 80 per cent of household heads were small scale part-time farmers by the mid-1980s (Kovach, 1991). Szelényi (1988) argued that having a wife and children at home made it more likely that a villager would have a small private plot for subsistence production. He defined woman's entrepreneurial role as purely passive although in one of his case studies he noted that 'their only piece of luck [in their attempt to run a private dairy farm] was that Laci's [female] companion had had some experience with cows' (Szelényi, 1988, p. 111) and 'Kerekes's mother . . . may be the most entrepreneurial, the most addicted to gambling of all. She pushed her son all the way . . . The daughter of a burgher-entrepreneur . . . probably she . . . never accepted the proletarian existence that collectivization offered' (Szelényi, 1988, p. 107). Despite his own field evidence he adhered to the patriarchal-paternalistic view of official socialism.

Post-Communist Rural Employment

Commuters from rural villages were forced to seek alternative employment nearer their homes as transportation costs increased. Privatization of agricultural land, closure of agricultural cooperatives and the loss of jobs in the towns led to high rates of unemployment in rural areas. Another factor was the collapse of the small industrial branch plants in rural areas. During privatization these plants were the first to go bankrupt. There was a choice between migration to the cities where living conditions were more expensive, or forced entrepreneurship in the villages, or to put it more positively, the challenge of economic independence. 'There is nothing else to do and we have a house we built here. I cannot commute to elsewhere because the village is too isolated. I would prefer a job' (entrepreneur in eastern border region village). In some areas the emergence of a bottom-up economic policy for rural areas including the growth of rural tourism has brought new economic vigour to these areas (Somogyi, 1999).

Rural areas have certain advantages for entrepreneurial activities. The 'push' factors of unemployment and lack of alternative opportunities are often more marked than in urban areas and the positive element of community and family networks is stronger. Women, as in most cultures, are most likely to be involved in maintaining these networks and in preserving traditions. The privatization of the collectives has to some extent released capital, equipment and agricultural resources in rural areas (Kovács, 1996). Many women were employed by the collectives in small manufacturing branch plants, or in services such as childcare or retail stores or bookkeeping. These skills can now be transferred to self-employment in the private sector. The availability of shops is more limited in rural than in urban areas and, as commuting to towns is reduced by the rising cost of transport, there is an increasing need for local services such as convenience and clothing stores, hairdressers, and restaurants. These services are particularly important to women who are least likely to have access to private means of transport and it is logical that women entrepreneurs should respond to these needs. They are also involved in providing secretarial and financial services in small family-owned businesses.

In the post-transition decade unemployment in rural areas was lower than in urban areas for women but not for men. For women it was lower than the national average but for men it was above the national average (Table 3.3a). However, for women economic activity rates were much lower in villages (43.9 per cent) than in towns (57.2 per cent) while for men the difference was only 7 per cent (58.9 per cent versus 65.6 per cent) (Asztalos Morell, 1999b: Table 1). Unemployment in rural areas may be under recorded, especially for women, in that the cost of travelling to the employment office in the nearby town is often greater than the financial support available (Timár and Velkey, 1998).

Overall, it appears that rural areas have shouldered the greater part of the burden of economic restructuring since 1990 because of the reduction of government agricultural subsidies, the break-up of cooperative and state farms and the closure of the industrial enterprises associated with them, and the loss of subsidized transport for commuters to nearby towns. Women in particular have become spatially entrapped in the countryside and many villages have a high proportion of elderly women. These women are generally surviving on inadequate state pensions sometimes eked out by small-scale subsistence production on small plots of land bought with compensation vouchers.

In the early 1990s, Hungary had one of the highest ratios of women to men (1.10) in East Central Europe but a lower percentage of women in the labour force (44.5) than in most countries in the region (United Nations, 1995). By 2004, higher female life expectancy was reflected in an even higher female to male ratio of 1.11 but this female dominance was much less in the villages, especially in the younger age groups (Table 3.4). The urban/rural difference may be a result of the better job market for women in urban areas, especially in the service and professional sectors. Hungarian women on average live almost nine years longer than men which may partly explain the preponderance of elderly women in many villages (Table 3.4). This life expectancy difference is greater than in most Eastern European countries but smaller than in Russia. Rural population density fell from 78 persons per square kilometre of arable land in 1990 to 77 in 1997 (World Bank, 2000). Early in the transition the Hungarian population also had the highest mean years of schooling in the region at 9.8 years in 1992. On a global scale, Hungary had the second slowest growing population in the world at –0.4 per cent between 1984 and 1995, the fifth slowest growing between 1995 and 2000 (–0.5 per cent) and was expected to be seventh on this scale between 2000 and 2005 with a continuation of the same rate of decline (Economist, 2002). How far this socio-economic profile influences gender and regional variation in the impact of the privatization, will be explored.

Definition of Rural Areas

Defining rural areas is always difficult (Bock 2004a) but in Hungary it is particularly complicated. Villages and towns are distinguished according to their system of local government rather than by population size. Thus it is possible to have villages with larger populations than towns. The rural population is usually considered to be all those not living in towns. Although the proportion of the population living in rural areas fell steadily from 62 per cent in 1946 to 41.9 per cent in 1985, the trend then reversed and by 1997 it was estimated that 47 per cent of the Hungarian population

Table 3.3a Gender differences in economic activity and unemployment rates for the rural and urban population, ages 15–74, in Hungary, 1987, 1990, 1992 and 1994

Date	Economic Activity Rate*			Unemployment Rate		
	Men	Women	Total	Men	Women	Total
Villages						
1987	71.5	54.7	63.0	–	–	–
1990	70.1	53.8	61.8	3.5	1.8	2.8
1992	60.8	51.9	56.2	18.1	9.1	13.8
1994	58.9	43.9	51.0	24.6	11.4	18.6
Towns						
1987	73.4	65.5	69.3	–	–	–
1992	68.9	60.5	64.3	11.4	9.1	10.2
1994	65.6	57.2	64.3	15.3	11.3	13.2
Hungary						
1987	72.5	60.7	66.5	–	–	–
1992	65.6	57.2	61.1	13.9	9.1	11.5
1994	62.8	51.0	56.4	19.0	11.3	15.2

*Recipients of GYES and GYED (childcare subsidies) are counted as economically active for 1987, 1990 and 1992. In 1994 only those recipients reporting secure employment upon return from childcare leave are counted as active.

Source: Asztalos I. Morell (1999b).

Table 3.3b Male and female unemployment, 1994 and 1998 for villages and towns in Hungary

Sex of Unemployed	Proportion of unemployed living in villages	Male/Female unemployed in villages	Male/Female unemployed in towns (except Budapest)	Hungary
1994				
Female	42.7	39.2	41.9	41.5
Male	47.1	60.8	58.1	58.5
Total	45.2	100.0	100.0	100.0
1998				
Female	42.6	42.0	46.6	45.2
Male	48.6	58.0	53.4	54.8
Total	45.9	100.0	100.0	100.0

Source: Bódi, F. and Cs. Obádovics (2000) Munkanélküliség a vidéki Magyarországon (Unemployment in Hungary). *Területi Statisztika* 3 (40) 1 January.

Table 3.4   Female to male ratio for settlement size and study areas in January 2004

Unit	Age Group								Total
	under 15	15–19	20–29	30–39	40–49	50–59	60–69	Over 60	
Győr-Moson-Sopron	0.96	0.93	0.93	0.94	1.02	1.07	1.32	1.73	1.07
Békés County	0.96	0.98	0.93	0.96	1.02	1.08	1.32	1.67	1.09
All villages	0.95	0.92	0.91	0.94	0.94	1.03	1.36	1.87	1.06
Hungary	0.95	0.96	0.96	0.98	1.05	1.14	1.37	1.86	1.11

Source: Hungarian Central Statistical Office, 2004.

lived in villages and towns of less than 10 000 (Kovács, 1997). In 2003 Győr-Moson-Sopron County in western Hungary had 43.5 per cent of its population living in villages or 46.1 per cent if we include towns of less than 10 000. In the south east, in the county of Békés, 35 per cent of the population lived in villages and a further 9.7 per cent in eight small towns (Central Statistical Office, 2003).

Even those living in urban areas had close links with rural life. Hungary is unusual in that counterurbanization occurred during the socialist period when, as a result of economic reforms, it became profitable to earn extra money from a household plot, and housing shortages made living in the countryside attractive (Paul, 1992). Many rural residents became peasant-labourers working in the urban-based state industrial sector during the day and on their land in their spare time. Swain (1996) calls this stage underurbanization in which villages were starved of infrastructural resources. In this way most Hungarians became familiar with life in both urban and rural areas. As a result of these rural-urban links, both the closure of state-owned industries in the cities and the break-up of the collective and co-operative farms in the countryside had a devastating effect on employment among rural dwellers.

The Story of One Village on the Western Border

However, foreign firms are moving on to greenfield sites, often on former cooperative lands, and setting up new businesses in the western border area. The mayor of Jánossomorja (in 2004 Jánossomorja became officially a town) told us in June 1998 that four foreign agri-business food factories and an electronics firm were employing about 700 people and a building materials plant was being set up and a gravel pit opened. Many of these new firms were built on land belonging to the former cooperative. In 1990 the village had had 600 unemployed but by 1998 had 150 people commuting from neighbouring settlements to work in Jánossomorja. Three of the new firms provide joint bus services for workers so they can work three shifts. The work was mostly unskilled and in the food firms 60 per cent of the workers were women as they were only required to be 'efficient with their hands, cope with a monotonous job and be prepared to work on a production line' (Jánossomorja, 25 June 1998). The village agricultural cooperative was active until 1990. It employed 650 people partly in livestock farming and crop cultivation and partly in the small ventures of the cooperative, and in a construction branch and a metalworking branch. Of these 650 workers only 120 have remained in employment on the cooperative. About 400 former cooperative workers have found jobs in the new local firms while about 100 have obtained work across the border in Austria. The men working in Austria were mostly steelworkers and carpenters while women go for seasonal agricultural work. Younger workers have official work permits but older workers, aged 40 to 60 years, do not. Some 9000 hectares of former state land is in foreign hands but 'nobody can really tell as there is no legal documentation of ownership transfer' as most of it involves 'pocket contracts' (Ibid).

The post-socialist counterurbanization trend reflects the improvement of village infrastructure undertaken by local governments in Hungary over the last decade, particularly in terms of roads, gas pipelines and new schools. In Jánossomorja the mayor told us that by 1998 a sewer system and gas supply had been completed, a solid waste refuse site established on 35 hectares of local government land, phone lines had become readily available, and new roads, pavements and parks had been constructed (Ibid). Many Austrians are buying houses in the village and housing construction has increased some 30 per cent (Ibid). A general realization that property prices and the cost of living are lower in the countryside, where it is possible to grow some of one's own food and to heat with cheap wood rather than expensive gas, as in the city, has further encouraged urban to rural migration. In Romania under socialism many villages were blighted by the threat of destruction. In one border village in western Romania most of the people were ethnic Germans who were able to migrate to Germany leaving behind empty houses. These houses were sold very cheaply by the Romanian government and so many became occupied by gypsy families but other people also moved from nearby towns. Some of these urban to rural migrants turn to self-employment in order to earn a living in their new rural residential locations. In Jánossomorja, the former German residents had left after the Second World War and were now living near Munich. Some of them came back to visit every year but the links were weakening. At the same time ethnic Hungarians from Czechoslovakia were settled in the empty houses left behind by the departed Germans. The village was emphasizing the teaching of both German and English in its schools and trying to maintain these external links (Ibid).

In the west of Hungary there is a recent history of private farming while in the east this was less widespread and most people remained as members of the agricultural proletariat working on the cooperatives although a few had become independent producers before the transition. It was these early entrepreneurs who had become some of the most successful in the 1990s. In the west, in Győr-Moson-Sopron County, unemployment was only 3.4 per cent in 2003 but in Békés County the unemployment rate was 7.1 per cent (Hungarian Central Statistical Office, 2003). Unemployment rates among gypsies were very high. Despite these differences, which might be expected to influence willingness to start businesses, in both areas rural people are responding to demand for services and products from local residents and from the externally generated growth of travellers and tourists.

Rural women are reported to be less likely than their urban sisters to suffer from stress, as housing is often less cramped and family networks remained in place although female unemployment was higher and the chance of finding a second job lower in the villages (Corrin, 1992a). Also rural women may have become more independent as their husbands commuted to urban jobs. However, with the rise in the cost of transportation and the reduced frequency of public transport, commuting workers are being squeezed out of the urban labour market and Tóth (1992, p. 1041) believed that 'closed rural labor markets' would evolve. Not only have villagers lost their jobs and their urban incomes but they have also lost their ties to urban culture and way of life and find themselves stuck in a more traditional paternalistic rural way of life. There are also *nouveau riche* individuals in the villages. These people are predominantly those who were able to benefit from the break up of the collectives, often the former managers. One such was raising racehorses and travelled by helicopter to and from his remote ranch in eastern Hungary. Thus rural society is becoming increasingly socially and economically polarized and employment in border areas may be more dependent on links across the border than on jobs in nearby Hungarian towns.

The urban bias in planning in Hungary (Enyedi, 1989) has meant that services and infrastructure in rural areas were much inferior to those in the cities so there is scope for the development of local services such as hairdressers and village stores. In Békés County in 1992 there were 13 388 firms of which 11 305 were individual entrepreneurs most of whom had legalized their former activities in the second economy (Groen and Visser, 1993). Firms in Békés County had the lowest profitability amongst Hungarian counties, except for the small entrepreneurs of whom 65 per cent were making a profit (Ibid). Only Vas County in western Hungary had a higher proportion of successful entrepreneurs (Ibid). In one of the border villages in Békés County, in the summer of 1996, we met a woman who had opened a village general store and an ice-cream stand. She had previously worked for the store on the collective farm but when the cooperative was broken up she got a loan and built her own store in the village. Her husband had been forced to take early retirement from the fire service in a nearby town and so she had become the family breadwinner. She had also invested in additional land which she and her husband farmed, producing both for household consumption and for sale. Although she clearly enjoyed her new role, she complained about shortage of time and excused her activities by explaining that she was only doing it because they needed money so their two teenage daughters could train for good jobs.

Workplace Location

The spatial separation of home and work was one of the major impacts of the industrial revolution as craftwork was replaced by factory-based mass production. This was accompanied by increasing gender specialization and spatial separation of productive and reproductive work. In Hungary, the intertwining of housing and labour markets enhanced this separation as the state, unable to afford to provide housing near the new factories for workers shifting from agricultural to industrial work, encouraged people to stay in the villages by offering free or very cheap transportation between work and home. This subsidized transportation was one of the first benefits to disappear in the transition but was being revived in some rural areas (Jánossomorja interview, 1998). People in these former commuter villages became spatially entrapped as they could not afford to move to live in the towns and cities.

In many cases this isolation was more difficult for women as their reproductive responsibilities, especially in caring for children and elderly parents, made it even more difficult for them to move than for men. An extreme case of this was a young woman who had Austrian citizenship through her father, but who gave up her job in Vienna to move back to a village in the western border area of Hungary to take care of her ailing mother. However, the social, cultural and economic capital she had accumulated from her links with Austria allowed her to set up in business as a beautician in Hungary serving Austrian, not village, clients. These differing reactions to village living also illustrate the variation between the attitudes in eastern and western border villages.

In many capitalist countries the family farm is the one place that allows the home and workplace to be spatially congruent. The industrialization of agriculture through large socialist cooperatives, often serving several villages, introduced separation of home and workplace into agriculture. Under socialism the village agricultural cooperatives provided work for women in small industrial branch plants, childcare centres, cafeterias, retail stores, or as accountants more often than in agriculture. The skills learned in these occupations could often be translated into independent businesses after the collapse of the cooperatives and the privatization of most agricultural land, as seen in some of our case studies. Many women took over the running of former cooperative stores. Increasingly, women can now combine their reproductive and productive work in the same location through agritourism and value-added sales on the farm (Jervell and Momsen, 2001). Perhaps the most innovative and successful translation of a skill learned in the previous regime to a successful business found in our survey, was the woman who had trained as a cook and then set up as an entrepreneur mixing animal feed for pig producers in several villages.

Houses in villages are often spacious enough to permit the operation of a small business and over two-thirds (69 per cent) of our survey respondents in the western study area combined their home with their business premises (Table 3.5). Setting up a workplace in the home had the advantage of permitting the deduction of the cost of heating and lighting of the house as a business expense. However, this dominant combination of home and workplace was not true in the east. In the eastern border villages the most popular location for a business for both men and women was working in non-residential premises in the same village. We had expected women entrepreneurs to be more likely to work out of the home as such joint use of the home would maximize the flexibility of their time and ease of combining childcare and paid work. This was true in the east where 47 per cent of women entrepreneurs worked out of their homes as compared to 38 per cent of men but in the west only 64 per cent of women versus 74 per cent of men worked at home (Table 3.5). Part of this difference may be due to the type of work undertaken by men: in the east, they often worked in agriculture, especially intensive greenhouse vegetable production on the edge of the village; while in the west many men worked in car repair or manufacturing of small metal items which they undertook in workshops attached to their homes (Plates 6.1 and 6.2). In both east and west women often ran small hairdressing, or dressmaking businesses out of their homes, with some in the west working as home-based outworkers for clothing factories in nearby towns.

The relative importance of using other premises in eastern villages appears to reflect lower costs of land and buildings than in the west, the availability of state property for

Plate 3.1   Closed state farm in the eastern border region

Table 3.5 Location of businesses in east and west border areas of Hungary, 1998

| | East (N = 101) | | West (N = 248) | |
| | | Percentage | | |
Location	Women	Men	Women	Men
At home	47.2	37.5	64.0	74.0
In same village	49.1	41.7	32.0	19.5
In another village	0.0	4.1	2.4	3.3
In a nearby town	0.0	0.0	1.6	2.4
In several places	3.8	16.7	0.0	0.8
Total	100.0	100.0	100.0	100.0

Source: Fieldwork.

rent (Plate 3.1) and declining populations leaving empty houses, as across the border in Romania. In the aforementioned case study of a feed business, the feed was mixed in outbuildings built in the yard of the family house and they had waited until the old lady in the rundown house across the street died to acquire her house, knock it down and build a large modern piggery on the site. The choice of location was partly imposed by new European Union regulations which did not allow residences and pigsties on the same site. In another eastern village a very successful woman entrepreneur had set up a village store next to her home. The store had been their house and in 1990 they built a new home next door and turned the old house into a shop and storage area. She had recently bought two old houses in the village. She used one as a place where she could provide lunch for schoolchildren and the elderly and was renovating the other to provide tourist accommodation. She also operated general stores in two other villages nearby and was elected to the local government.

As Table 3.5 shows few entrepreneurs in either east or west worked in a village other than the one in which they lived. In the east, the post-World War II international border changes had cut off the villages from the towns which had previously been their market centres but in the west the towns and cities were on the Hungarian side of the border. These political and geographical differences are reflected in the location of rural businesses (Table 3.5). In both border regions men were more likely than women to work in a neighbouring village, usually one where they had family connections or links based on their workplace under communism. In the west a few people had set up businesses in a nearby town because of a larger customer base and relative proximity.

In some cases, low rural customer numbers had forced some entrepreneurs to utilize multiple sites. In the east, the small size of villages and low level of disposable income led many male entrepreneurs, 17 per cent of those surveyed, to become itinerant traders depending on sales in different periodic markets or door to door sales to maximize their customer range. A few entrepreneurs, especially women, had set up a second store or pub in a neighbouring village, usually one in which they or their husband had been born. In the west there were even a couple of city dwellers who supplied services to villages in which they had previously lived: for example a woman, now living in Sopron, owned a vineyard and made her own wine and sold this in a winebar in the village close by. One woman in the west had run a flower stall in the summer at Lake Balaton, Hungary's main tourist area and one in Sopron in the autumn but had recently given this up as she found the rent for the stalls too high. She now operates a flower and ice cream shop out of her home in the village. Several eastern entrepreneurs told us that they had given up working as itinerant salesmen as they were now undercut in periodic markets by Chinese immigrants. However, in both east and west border villages Romanians selling hand embroidered goods are still a common sight.

Conclusion

It appears that in the early stages of the transition rural women in Hungary were more likely to become entrepreneurs in poorer areas where the need for services was greatest and alternative opportunities scarce. In western Hungary opportunities for employment in Austria or in new local factories provided alternatives but eventually were seen by many women as more inconvenient than becoming an entrepreneur. In addition, as tourism expanded in the west many women were able to set up businesses serving these visitors. For many rural women such activity is still seen as a last resort when husbands are unable to support the family, or at least it is publicly justified in this way. Other women suggest that it was their husband's initiative that encouraged their own venture into self-employment. Thus traditional rural family values survive despite decades of state-sponsored gender equality now reinforced by global influences encouraging female employment. In general, a family background of entrepreneurship plus class and educational differences between women may be more important than local opportunities in encouraging women's entrepreneurship (Kovács and Váradi, 2000).

Chapter 4

Entrepreneurship in Rural Eastern Europe

Interest in entrepreneurial behaviour in Europe's peripheral rural areas is growing as the European Union moves from a focus on supporting agricultural production to recognizing that rural economic development is now more complex (Labriandis, 2004). Kalantaridis (2004, p. 62) notes that 'economic growth in rural areas is invariably conditioned by the pervasive influence of a myriad of often small and micro-scale entrepreneurial ventures'. Rural areas suffer from a shortage of potential entrepreneurs because of outmigration and so the building of a 'critical mass of entrepreneurship is a key economic development issue' (Ibid). So far we know little about 'how the characteristics of the rural influence the incidence and attributes of entrepreneurial agents, and probably, less about the function of entrepreneurship as an engine of growth and structural transformation in the countryside' (Ibid). Labrianidis et al. (2004) report on a survey of 996 innovative entrepreneurs in ten peripheral rural areas of five European countries. The results of their surveys throw new light on the relationship of entrepreneurial capacity to the growth of peripheral rural areas. In our study in Hungary we looked at two areas which had long been peripheral but in which the changing status of the international borders is currently either reinforcing or reducing this peripherality. Testing our findings against those of the Labrianidis (2004) study is helpful in assessing the importance of these border changes to the growth of entrepreneurship.

The literature identifies two key research issues which are particularly relevant to our study: what drives individuals to become involved in entrepreneurial pursuits, and whether spatial factors influence their decisions (Kalantaridis, 2004). Some of the negative factors specific to rural areas are small markets and in the case of transition countries, imperfect markets (Ibid). This is where cross-border trade becomes important. However, production factor endowments such as human capital, land and buildings may be positive for entrepreneurs in rural areas. The Labrianidis (2004) study of entrepreneurship in ten areas in Europe revealed that differences were greater between countries than within countries but at the same time locality mattered. Thus by focusing on two areas within one country we are minimizing national differences and maximizing local differences. The Labrianidis study (2004) derived indices of 12 entrepreneurial characteristics of individuals for each case study area, by comparing entrepreneurs to the general population in the area. These indices covered gender, education, managerial background, previous enterprises, in-migration, manual background, family influence, unemployment before start-up, administrative background, university education and age. We also included most of these aspects in our surveys although we had less on an individual's entrepreneurship history, given the limited possibilities for self-employment in the past. However, we did find that quite a few of our post-socialist entrepreneurs, or their immediate family, had relevant past experience. We also included questions on family size and the availability of childcare as we believed these impacted on women entrepreneurs in particular. Our findings will be compared to those of the Labrianidis survey in Chapter 5.

Entrepreneurship in Hungary

Entrepreneurship is at the heart of the post-socialism changes in Eastern Europe (UNECE, 2002). It is an important source of job creation and new career opportunities for both men and women. However, access to entrepreneurship is not gender neutral but reflects asymmetries in the broader labour market. The transition to capitalism in East-Central Europe (ECE) has been accompanied by an employment crisis and a movement into self-employment (Smith, 2000). It has been estimated that there were 26 million fewer jobs in the 27 countries of East Central Europe, and the former Soviet Union in 1997 relative to 1989 (Unicef, 1999). Of these about 14 million were jobs lost by women so it is clear that women coped with a high proportion of the costs of adjustment. As a result, self-employment became for many women the only path for paid employment. However, women entrepreneurs faced gender specific barriers in access to information, networks and collateral (UNECE, 2002).

A variety of household responses to the employment crisis emerged after 1989. In Hungary, as income from wages declined, there was a growing dependence on self-employment and state pensions seen by Smith (2000, p. 1770) as 'a combination of household and individual resilience, and reliance on state redistribution'. During the 1990s atypical work contracts began to appear in conjunction with the expansion of the private sector, enterprise restructuring and expansion of the parallel economy (UNECE, 2002, p. 16). The parallel economy included activities which were legal, though not registered, and illegal and/or criminal activity such as prostitution and drug trafficking. The expansion of these activities was a response to high social security contributions in the formal labour market which many firms could not afford, and to institutional weaknesses, including lack of law enforcement, which allowed for high profits from illegal operations. Kalantaridis (2004) notes that a lack of legitimacy for entrepreneurial pursuits in post-socialist countries often results in a focus on informal or outright illegal activities in these areas. This is especially a problem in border areas where criminal activity includes smuggling of people and drugs, and employment of illegal immigrants. In the western border area of Hungary this largely involves women brought in from the Ukraine, Belarus and Romania to work as prostitutes, while on the eastern border in addition to foreign prostitutes, Romanians of both sexes are present as traders and agricultural workers. In our survey we found that some entrepreneurial activities were not officially registered, as the laws requiring such registration changed over time.

The Development of Entrepreneurship

In the post-socialist countries of Eastern Europe, Brezinski and Fritsch (1996) suggest that bottom-up development based on small new firms may make a considerable contribution to the solution of social, economic and political problems. This approach is welcomed by the new governments of the region as money is no longer available to subsidize large state-owned enterprises and top-down privatization is proving more difficult than expected. Bottom-up development is thought to fulfil four functions in the process of transformation to the market economy: firstly it creates a pluralistic society and safeguards the existence of democratic society; secondly it contributes to competition and may help protect the market from the collapse of dominant large companies; thirdly small firms improve the flexibility of the economy as a whole by encouraging the adoption of innovations, adaptation to changes in supply and demand and reduction of regional disparities; and fourthly small firms are seen as providing a social cushion by absorbing some of the job losses caused by the collapse of state enterprises (Brezinski and Fritsch, 1996, pp. 3–4).

The success of new, small enterprises depends very much on the economic environment at the starting point. In Hungary 'quite widespread forms of artisan/craft based small scale private production were tolerated throughout most of the post war period' (Bartlett and Hoggett, 1996, p. 154). The limit to the number of employees permitted in such enterprises was gradually raised starting in 1982 and by 1986 Hungary had over 36 000 small enterprises (Bartlett and Hoggett, 1996, Table 11.2). This was in addition to the widespread private production of food from small family plots. This 'second or black' economy was more openly tolerated in Hungary than in most of the other countries under communist regimes. The inter-household and community networks upon which it was based now provide a crucial means of acquiring the skills, information and capital necessary to start up new enterprises in either the formal or informal sectors.

Such networks are especially important in Hungary where the official enabling structure for raising capital is a major deterrent with high interest rates varying between 30 per cent in 1989 to a low of 21 per cent in 1992 and rising again to 28 per cent in 1995 (Barclays, 1996), and a banking system which for a long time refused to accept property as security for loans (Bartlett and Hoggett, 1996, p. 156). A woman entrepreneur in an eastern Hungarian village interviewed in August 1996 was paying 35 per cent interest on her start-up loan. It is not surprising that most entrepreneurs interviewed in 1997 and 1998 went to great lengths to avoid taking out loans. They preferred to depend on family sources or earned start-up capital themselves by working abroad.

Bartlett and Hoggett (1996) consider that the mechanism for state encouragement of small business development in Hungary is pretty rudimentary and the World Bank (1995) is encouraging further structural reforms. It has been found that, in general, attempts to support entrepreneurship through capitalization of unemployment benefits and provision of technical assistance to the unemployed through local labour offices, have not been very successful (Fretwell and Jackman, 1994, p. 184; Laczló, 1996; Voszke, 1995). One of the main driving forces for entrepreneurial activity today is the tax benefits of self-employment, but the tax code is complex and frequently changing. Despite this, the proportion of small firms, with fewer than 50 employees but excluding sole proprietors, legally registered rose from 55 per cent of all enterprises in 1989 to 88 per cent in 1992 (Bartlett and Hoggett, 1996, Table 11.7). It is estimated that the number of sole proprietorships increased from 221 794 in 1989 to 339 866 in 1991 but the majority of these operate on a part-time or second-job basis and have no legal status (Bartlett and Hoggett, 1996, p. 164). Over one-fifth (21 per cent) of these small firms are in the trade sector in Hungary (Bartlett and Hoggett, 1996, p. 166) responding to the breakdown of the centrally planned system of retail distribution. The European Bank for Reconstruction and Development considers that Hungary is short of shops and there is much enthusiasm for investment in this sector (Bogyó, 1995).

Research in the United Kingdom (Curran and Blackburn, 1991) has shown that entrepreneurial activity is often a response to unemployment and low wages and to limited unemployment benefits. In Hungary where such benefits cover only 78 per cent of those looking for work (Bartlett and Hoggett, 1996, p. 154) it is not surprising to find a ready supply of new entrepreneurs. In Western Europe it has been shown that only 2–3 per cent of the unemployed become successful entrepreneurs but this option may have greater significance in Eastern Europe because of the limited development of small businesses (Fretwell and Jackman, 1994, p. 183). Our fieldwork revealed that in rural areas many of the women entrepreneurs interviewed were responding to a household situation in which husbands had lost their jobs or been forced to go on a disability pension. Curran and Blackburn (1991) suggest that those who experience discrimination in the labour market, such as ethnic minorities and women, frequently seek a solution in self-employment.

The 1995 UN *Human Development Report* stated that the objective of development is to enlarge people's choices. This involves equality of opportunity for all and the promotion of gender equality and other human rights. To make choices people have to see themselves as free to make decisions and to have individual agency. Under communism not only was independent action discouraged, but people felt that they did not need to think and worry about their future as the state would tell them what to do and take care of them. In agriculture the state farms absorbed the risk for farmers. Thus the loss of this umbrella of state protection with the transition has had a psychological effect, leaving many people not knowing how to make decisions and frightened by the concept of individual responsibility.

This psychological impact of the transition may provide another reason for women undertaking entrepreneurial activity or agreeing at least to register as an entrepreneur. As such work is seen as inherently risky, it is socially more acceptable for a woman to lose face by failure than for a man. Women may feel that if their business is not successful they can move back into reproductive work and a dependence on state childcare subsidies, and out of the formal labour force. On the other hand, many women find that being self-employed provides the time flexibility not otherwise available because of the lack of part-time jobs. It appears also that there are often tax advantages for the woman to be the officially registered entrepreneur. Women entrepreneurs were concentrated in retailing and personal services because these sectors were easy to enter as they required relatively low capital investment (Hisrich and Fülöp, 1997). Sometimes the husband might be the official entrepreneur but wives often provide a range of hidden and unpaid services without which the business might not survive (Nagy, 1997). Women rural entrepreneurs had few links with broader civil society. Women's groups are concentrated in Budapest and there seemed very little awareness of or identification with such groups by rural businesswomen, although one of our women entrepreneurs in eastern Hungary told us that her son-in-law's mother was the head of the women's group in Békéscsaba. Ideas of personal empowerment or individual achievement were very rarely openly expressed. Most rural women entrepreneurs saw themselves as helping their families and their community rather than trying to improve themselves individually.

Many writers consider the collapse of communism in Eastern Europe and the Soviet Union in 1989–90 to be a cataclysmic break, 'the end of History' (Fukuyama, 1992). The development of an entrepreneurial spirit is seen as a new beginning but we argue that, at least in the case of Hungary, it has deep roots in the past. The importance of a family history of entrepreneurship was also seen as an important characteristic of innovative entrepreneurs in the Labrianidis et al. (2004) study. However, the legacies approach (Szelényi, 1988; Kuczi, 1998) focusing on pre-socialist era family history and more recent experience in the second economy, plus Party membership, has declining utility as an explanation for either gender or regional differences in current entrepreneurial activity (Szelényi, 2001).

Hungary has a very high rate of entrepreneurial activity, accounting, in 1995, for slightly over one-fifth of employment in non-agricultural sectors, more than twice the average for OECD countries (11 per cent) and almost the average level for Latin America (26 per cent) (Gábor, 1997, p. 164). Gábor (1997, p. 159) sees this rapid growth of entrepreneurial activity since 1989 as 'too many, too small' and he further argues that the development of successful small businesses has been hampered by the historical heritage of the second economy as well as by the context of the crisis of the transition.

Entrepreneurs did exist during the communist period in Hungary but to a much smaller extent because of mixed messages from the state. The failure of the state to provide housing led to an expansion of privately built housing, especially in villages. Between 1971 and 1975, 44.2 per cent of all homes were built by the private sector (Róna-Tas, 1997, p. 111). The Politburo in both 1968 and 1976 had allowed small private artisans to survive in order to meet some of the demand for repairs to faulty consumer durables, and to undertake plumbing and electrical services (Róna-Tas, 1997, p. 106) but in the early 1970s the level of official criticisms and restrictions rose again. However, '[B]y 1976, the increasing discontent in the countryside over the lack of services and the ghost of the burgeoning shadow economy had persuaded the leadership to adopt a more favourable position toward private services in small towns and villages' (Róna-Tas, 1997, p. 111). Thus there was a more extensive history of entrepreneurial activity in rural than in urban areas.

Róna-Tas (1997) points out that, based on data from the 1981 social stratification survey, 70 per cent of the full-time self-employed were men, as women were too busy with their double burden of a full-time job in the state sector combined with their household chores. This was despite the fact that Hungarian women's workload had fallen from an average of nearly 75 hours a week in 1965 to 63 hours a week in 1977 which was still about ten hours a week more than the average for British women in 1961 (Unicef, 1999, p. 25). Women entrepreneurs were relatively most often involved in household farming (41 per cent) and in property letting (55 per cent) in this period. Men were also more likely to have the requisite skills for the types of work available, particularly repair of consumer durables such as televisions and cars, and construction work.

Many people in the 1960s and 1970s moved back and forth between the state and self-employment. In a 1982 survey of 389 self-employed artisans and tradesmen in eastern Hungary reported in Róna-Tas (1997) only 17 per cent said that they had chosen self-employment for material gain, and 14 per cent because of conflicts at their former place of work. Interestingly, half said it was because they thought it was a way of realizing their dreams, suggesting an unfulfilled desire for self-employment. It was generally believed that having a particular skill or training was the most important determinant of success. Given the acute consumer shortages of the period there was little competition so marketing strategies were unnecessary. Today in the post-communist economy, many entrepreneurs, especially in eastern Hungary, complain about having to deal with competition from other small businesses.

In 1982 laws came into operation which allowed an increase in the number of people who could work in private partnerships, permitted the private sector to do business with the state sector and liberalized the operating conditions for small family-owned firms. This led to an expansion of part-time work especially in internal contracting and outside consulting for state firms. These changes did not lead to an increase in hours worked but rather, as Róna-Tas (1997) points out, a redistribution of effort from the state to the private sector. Unlike the earlier reforms, these brought into the private sector those who had benefited most from the socialist economy. Consequently, when in 1985 at the Party Congress, the leadership tried to tighten its hold on the economy and restrict the private sector, the delegates baulked. The boundaries between public and private sectors became blurred and the 'private sector of needs' was transformed into a 'private sector of opportunities' weakening the power of the Party-state (Róna-Tas, 1997, p. 163). According to official national statistics, there has been a very rapid growth in the number of small entrepreneurs from 113 000 in 1980 to 211 536 in 1988, 394 000 in 1990 and 521 000 in 1991 (Róna-Tas, 1994, p. 50 for 1980 and 1988; Lengyel and Tóth, 1994, p.13 for 1990 and 1991). Lengyel and Tóth (1994) show that this growth in actual entrepreneurs was coupled with a substantial rise in the proportion of working age people who could envisage setting up business on their own account from 25 per cent in 1988 to 40 per cent in 1990. There was a slight change of definition in the reported national statistics in 1992, and as a result the number of enterprises without legal entity grew from 676 804 in 1992 to a peak of 936 312 in 1995 with the subset of sole proprietors falling from 89.6 per cent in 1992 to 84.5 per cent of this total in 1995. By 1997 the total figure had fallen to 839 602 of which sole proprietors made up 78.6 per cent (World Bank, 1999, p. 106). The decline in the proportion of sole proprietors indicates a growth in the size of enterprises.

Such a change of individual actions and aspirations reflected changes in official attitudes to small business. The laws that came into effect during this period enlarged the scope for entrepreneurship. Law No. 6 of 1988 on Business Societies, Associations, Companies and Ventures provided for economic associations. Law No. 13 of 1989 on Transformation of Economic Organization and Business Associations provided for the transformation of state enterprises. Law No. 5 of 1990 simplified the conditions for launching individual ventures, served as the legal protection of private property and promoted unbiased competition. The concepts of enterprise and entrepreneur were no longer vilified but became widely known and positively reported on by the mass media. At the same time the existing economy was in crisis and the formerly unknown condition of unemployment was becoming a reality (Lengyel and Tóth, 1994). Thus improvement in the conditions for self-employment and the need for alternative sources of income began to reinforce each other.

The development of a politically conscious entrepreneurial class occurred very rapidly after 1989 (Agócs and Agócs, 1993). At the end of 1991 there were over

400 000 individual entrepreneurs and 10 000 private firms active in Hungary and six months later the total number of enterprises had reached 638 275 (Agócs and Agócs, 1993).

Gender Differences in Labour Market Changes

Gender differences were seen in employment cuts, sectoral changes in employment and access to jobs in the private sector. On the whole, women did not benefit as often as men from job openings in the private sector and in the more dynamic branches of the service industry, largely due to increased gender discrimination in the labour market.

Unemployment

In the early stages of the transition for most countries of Eastern Europe, with the exception of Hungary, women were proportionately more likely to be unemployed than men and to remain out of work for longer (Heinen, 1994; Ciechcińska,1993; Timár, 1995; Micklewright and Nagy, 1996). It has been argued that the unique position of Hungary in terms of gender differences in unemployment may be the result of the early collapse of heavy industry which mainly employed men (USAID, 1991) but this does not explain the continuation of this differential. In Hungary, a country of ten million people, one-third of women's jobs have gone since 1989 (Unicef, 1999). A recent World Bank report suggests that two additional factors may explain this unusual feature of the Hungarian labour market: first the higher proportion of women in the government sector, which provides greater job security; and second, the higher proportion of women with college and university education relative to men in the labour market (World Bank, 1999, p. 232). Women made up about two-fifths of the unemployed in 1990. Figures for 1994 showed women as forming 41.8 per cent of the unemployed nationally but 51.5 per cent in Budapest. In 1996 official statistics show that women still had a lower unemployment rate than men in every county except Budapest (Central Statistical Office, 1998). A detailed study of one county, Békés (Timár and Velkey, 1998), reveals that settlement size also affects gender differences in unemployment rates and many women, although technically not unemployed were on maternity or childcare leave and so supposedly economically inactive.

But by 1997 only five of the 11 transition countries had higher female unemployment than male. In Hungary in 1998 the ratio of female to male unemployment was the lowest among the eight EU accession countries. It had fallen further by 2001 to about 78 per cent of male unemployment, although Lithuanian female unemployment had fallen more rapidly to 68 per cent (UNECE, 2002, Chart 1). In many countries, lower unemployment rates reflected women's withdrawal from the labour market, especially in the first phase of the transition process (UNECE, 2002, p. 9; Smith, 2000). This was no longer true after 1997 in Hungary as women's economic activity rates increased more than men's (Ibid). Between 1985 and 1997 male activity rates fell by 13.5 per cent and female rates by 18.5 per cent but between 1997 and 2000 female rates increased by 3.0 per cent while male rates only grew by 1.5 per cent. Consequently, male economic activity rates, for the working age population aged between 15 and 74, which had been 73.9 per cent in 1985 had fallen to 61.9 per cent in 2000, while the female activity rate had decreased from 61.3 per cent to 45.8 per cent over the same period (UNECE, 2002, p. 13).

Part-time and Atypical Forms of Employment

Self-employment may offer the flexibility needed by many women in order to combine their productive and reproductive work. The part-time work undertaken by young mothers in much of Western Europe is generally not available in these transitional economies, although desired by many (Table 4.1). A 1988 study found that 77 per cent of Hungarian working women wished to continue with their jobs even if given the opportunity to stay at home although ideally they would prefer part-time work (USAID, 1991, p. 57). By 1995 one-third of Hungarian mothers of young children wanted to stay at home and 56 per cent of fathers thought mothers should stay home if the husband could provide an adequate income for the family. However, 51 per cent of the mothers wanted to work part-time, although only just over one-third of husbands wanted their wives to work part-time (Table 4.1). When asked if they wanted a career only 43 per cent of women said they did want their own career (Table 4.1). Clearly, the flexibility of entrepreneurship meets many of these expressed wishes, especially of those husbands who wanted their wives to stay at home. In 2001 the proportion of women in part-time work in Hungary was only exceeded by that in the Czech Republic and Estonia (UNECE, 2002, p. 17). Women held about two-thirds of all part-time jobs in Hungary in 2001 (Ibid). It is not easy to interpret these figures as in our fieldwork we found both men and women who held both a full-time job and a part-time job.

Table 4.1   Desired type of work and interest in a career among Hungarian women, in percentages

Type of work desired if income not needed	Mothers	Fathers	
Stay at home	33.0	56.3	
Part-time work	51.1	35.3	
Full-time work	13.6	6.0	
Don't know	2.3	2.4	
Totals	100.0	100.0	
Women's interest in having a career	*Yes*	*No*	*Total*
Full-time workers	7.1	7.1	14.2
Part-time workers	24.6	28.2	52.8
Stay at home women	11.4	21.6	33.0
Totals	43.1	56.9	100.0

Source: Pongrácz, Tiborné and S. Molnár (eds) (1995), *To Raise a Child*, Budapest.

Employment cuts were deepest in the first half of the 1990s, when GDP sharply declined and countries were forced to introduce austerity measures which involved the closure of many firms that were no longer profitable under the new economic circumstances. Between 1991 and 1994 employment fell by 26.3 per cent for women and 25.5 per cent for men in Hungary while in the neighbouring countries of Poland, the Czech Republic, Slovakia, and Romania the decline in employment was greater for men than for women (UNECE, 2002). For the period 1995 to 1998 the loss of jobs slowed but in Hungary women still lost more jobs than men (2.4 per cent against 0.6 per cent). Over this same period, women in Poland and Slovakia gained more jobs than men (Ibid). However, between 1999 and 2000 women in Hungary gained more jobs (4.3 per cent) than men (3.0 per cent) whilst the other new countries of the European Union such as Poland, the Czech Republic, Slovakia and Slovenia showed continuing declines in employment for both men and women (Ibid).

Women and men in Hungary have long had equal access to education and recently women have often outnumbered men in tertiary education (Hrubos, 1994) (Table 5.4). Wong (1995) and Wong and Hauser (1992) show that in Hungary, Poland and Czechoslovakia there were distinct cross-national and gender differences in job mobility. These differences persisted despite policies of equality of opportunity in access to education. In Poland, Czechoslovakia and Hungary secondary education was divided into technical and humanities sectors. Boys were concentrated in the technical sector and so completed some form of vocational training as part of their secondary education. Thus they entered the labour market as skilled workers while girls, despite higher levels of general education, were considered to be unskilled (Corrin, 1994). The few women who qualified for shop floor work often chose to move out of such jobs because of the unpleasant working conditions and sexual harassment (Kürti, 1990). During the transition the level of education was an important factor behind raising wage disparities in Eastern Europe (Rutkowski, 1999) but there is little evidence that women have benefited from these increasing returns to education (UNECE, 2002).

The gender dimension of restructuring is seen most clearly in the changes to women's status as workers (see Table 4.1). 'In a region of the world which once enjoyed the distinction of having the highest rates of female labour force participation and the largest female share of paid employment, women now face unemployment, marginalization from the productive process, and loss of previous benefits and forms of social security' (Moghadam, 1992, p. 18; Habuda, 1995; Koncz, 1995). Sziráczki and Windell (1992), in a comparative study of Hungary and Bulgaria, showed that when recruiting skilled workers 66 per cent of establishments reported that they preferred to hire men while only 15 per cent had a preference for women. In Bulgaria 54 per cent of firms had a preference for men and 25 per cent for women while in Hungary the highest reported preference for men was in the state sector (71 per cent) while in Bulgaria it was in joint stock companies (64 per cent). In Hungary there was little open discrimination against women in recruiting professional staff and women dominate numerically in Hungarian banking and finance. There is increased use of gender specific job advertisements but only in Poland are women directly discriminated against as 'women' or 'mothers' (Ciechocińska, 1993).

Women face labour market vulnerability based on a human-capital based interpretation of the reserve army hypothesis. Women lack, or are unable to capitalize on, certain resources that are useful for retaining employment, such as previous Communist Party membership or experience of self-employment. Maternity leave and childcare made it seem that women were less ambitious and reliable.

Occupational sex segregation today is influenced by women's revalued resources. Women are more likely to have had an academic education rather than vocational training and to have had experience in the service sector. The service sector has been the fastest growing sector since transition while heavy industry has collapsed. The hospitality industry was one of the first to be privatized and offers opportunities for self-employment for women, as does the rapidly expanding retailing sector. Professional women's qualifications are also being revalued. In 1980 almost half of CFOs and chief accountants in Hungary were women. These leadership posts had little prestige and involved relatively low levels of authority and autonomy in economies where most financial resources were centrally allocated. Today, foreign banks and multinational accountants turn to these trained women, especially if they also speak several languages, for high-paying positions and young men are increasingly enrolling in business courses at the Budapest University of Economics. Salaries in these professions are now close to those of Western Europe and four to five times those in teaching. However, even for highly educated young women accountants speaking several languages, gender discrimination can still make it hard to get the best jobs (Zs. Szörényi, personal communication, 2000).

During the restructuring process in Eastern Europe there was a shift of employment from agriculture and industry to services, as had occurred earlier in Western Europe (Table 4.2). Women were disproportionately affected by this shift with marked declines in their involvement in agriculture except in Latvia, Estonia, Slovenia and Slovakia and in industry throughout the region. The decline in jobs in industry reflected a shedding of clerical jobs and restructuring of light industries, such as textiles, and the closure of many of the small workshops associated with agricultural cooperatives.

Table 4.2 **Percentage share of women in total employment by industry in Hungary, 1992 and 1996**

Economic sector	1992	1996
Agriculture	31.2	24.8
Industry	41.3	38.6
Total services	54.2	52.6
Financial intermediation	76.0	66.3
Real estate, renting, etc.	51.2	46.8
Public administration	34.7	42.7
Education	75.8	76.1
Health and social care	75.1	75.3
Total for all sectors	**45.7**	**44.0**

Source: UNECE *Economic Survey of Europe 1999*, No. 1. Based on national labour force surveys. Adapted from UNECE, 2002, p. 11.

Changes in the employment structure within the service sector (see Table 4.2) reveal that women moved into public administration, teaching and health services. At the same time women's share of employment in financial services declined. In Hungary, women had long dominated dentistry, as this sector of the medical services was seen as having less status than that of medical doctors. However, after the transition dentistry became profitable in the western border area serving dental tourists from neighbouring Western European countries. It is said that Sopron, close to the Austrian border, has more dentists per capita than any other town in Hungary. The result has been that the long-standing female majority in dental schools has been replaced by a majority of male students (I. Szelényi, personal communication, 2001). On the whole, women moved into public services and benefited less than men from the expansion of market driven services. These changes represent a move to lower paid jobs: between 1993 and 1997 wages in the financial sector rose from 190 to 199, and those in education fell from 98 to 86 relative to an average wage of 100 (UNECE, 2002, p. 12). Prior to 1989 average female wages in Eastern Europe were 20 to 25 per cent lower than men's, largely due to occupational segregation (UNECE, 2002, p. 9). Under a market economy the further polarization of wages and the continuing gender gap in pay has had a negative impact on women's financial status.

Women Entrepreneurs

Barbara Einhorn (1993a, pp. 135–37) noted that '[S]elf-employment is much favoured by policy makers right across Europe, and in the case of former state socialist countries has the advantage of encouraging talents not rewarded in the past, namely enterprise, individual initiative, and resourcefulness . . . However, it would be naïve to imagine that this could have a major impact on female unemployment in the current democratization process. Rather it must remain a minority solution.' Kalantaridis (2004), reports that this situation for women entrepreneurs is true in all the areas surveyed. Minniti (2003, p. 4) found in a 2002 survey of 37 countries worldwide that women's participation rates were lower than those of men in every country with Japan being at the low end and Thailand at the high end and Hungary in the middle, with a higher proportion of women entrepreneurs than any other transition country surveyed. Nagy (1997) suggests that there is a higher proportion of women entrepreneurs in Hungary than in most countries in Western Europe and that the number of women entrepreneurs had increased rapidly after 1990 to 41 per cent of businesses. She found that 36 per cent of the businesses set up by women after 1990 were part-time and that the number of retired women entrepreneurs had decreased. Nagy (1997) sees this growth of self-employment as being an important part of household survival strategies, especially as a way of reducing taxes.

The 2003 survey (Minniti, 2003) showed that for opportunity entrepreneurship, where people choose to start their own business as one of several desirable career options, there are almost twice as many men as women (Ibid). However, for necessity entrepreneurship where people become self-employed because of lack of options, the proportions of men and women are about the same (Ibid). Nagy (1997) found that forced entrepreneurship was slightly more common for women than for men but that far fewer women chose willingly to become self-employed. Minniti (2003, p. 11) suggests that women's relatively high involvement in forced or necessity entrepreneurship may indicate that 'self-employment is used as a way to circumvent institutional and cultural constraints with respect to female employment, as well as a way to provide supplemental family income'.

Most published studies of Hungarian entrepreneurship deal with the situation in the very early years of the transition (see, for example, Agócs and Agócs, 1993). During the 1990s the gap between men and women in entrepreneurial activities widened in most Eastern European countries (UNECE, 2002). However, Koncz (2000) suggests that the small business sector is increasingly important for women. In 1988 women accounted for less than one-third of workers in full-time businesses in the private sector and this proportion rose to almost two-fifths by 1993. In this latter year, there were 213000 one-woman businesses in Hungary, accounting for more than 40 per cent of all sole proprietorships. A further 37000 female proprietors were registered in partnerships, representing 30 per cent of the total. Many women are in business only as a sideline activity, however, and only 37 per cent of full-time sole traders were women in 1993. Women accounted for 56 per cent of the sole proprietors on a pension. In 1999 Koncz (2000) estimated that of the 734000 small and medium-sized enterprises in Hungary, 38 per cent were run by women. Of these, 16 per cent of the companies and partnerships were registered to women, and 51 per cent of the sole proprietors were women. The International Labour Office reported that Hungary had the largest increase in the proportion of self-employed workers of any country for which data was available, between 1996 and 2000 (ILO, 2003). Men entrepreneurs increased from 15.7 per cent of the total number of workers to 18.8 per cent and women from 8.2 per cent to 9.5 per cent.

By 2000 there were twice as many businesses owned by men as by women in Hungary (UNECE, 2002, p. 15). This figure of one-third women-owned businesses is slightly below the level we found in our research in rural Hungary. It has been argued that the participation of women among entrepreneurs was relatively high in Hungary even before 1990 in comparison with the situation in Poland or the Czech Republic (UNECE, 2002, p. 24). Most women entrepreneurs in Eastern European countries work in small or micro-enterprises and are involved in activities such as retail services, cross-border trade, subcontracting work at home or street trade (UNECE, 2002, p. 14). In our research in Hungarian border villages we also found that women's businesses were very small but we noticed marked regional differences in their types of activities. In western Hungary subcontracting work at home was quite common while it was not recorded in the eastern border region. On the other hand, cross-border and street trading was more common in eastern Hungary than in the west.

Szabo (1992) reported that 32 per cent of the owners, managers and members of the new small enterprises created immediately after the fall of communism were women. Their participation rates varied by type of enterprise, being highest in trading enterprises at 65 per cent and making up 43 per cent in new small cooperative activities. The new privatized businesses made use of women's knowledge gained from the skills learned from management of the household economy, their involvement in private or semi-private spheres of production and their high levels of education and training especially in economics and finance (Szabó, 1992). Women's role in such enterprises has been encouraged partly because the hotel and catering sector, in which a lot of women are involved, was one of the first to be privatized in 1990 (Groen and

Visser, 1993). Lengyel and Tóth, in a representative survey of the adult population of Hungary covering 3000 individuals in 1988 and 1000 in 1990, found that women's interest in entrepreneurship increased from 16 per cent of those surveyed in 1988 to 37 per cent in 1990 while 37 per cent of men surveyed in 1988 and 54 per cent in 1990 had entrepreneurial inclinations (Lengyel and Tóth, 1994). Lazreg (1999, p. 35) reported that 51 per cent of sole proprietors were women. In general, the 1999 Unicef report found that women appear to be less inclined or less able to move into self-employment or entrepreneurship than men throughout the ECE and CIS countries, stating that further research into women's role in entrepreneurship is needed (Unicef, 1999, p. 40).

In the USA there has been a very rapid growth in the number of women-owned enterprises and they are the fastest growing business segment. In 2004 the Center for Women's Business Research said that there were 10.6 million privately held businesses in which women owned at least 50 per cent of the company, accounting for roughly half of all US businesses (Le, 2005). Between 1997 and 2004 the growth in women-owned firms was nearly double that of all US firms – 17.4 per cent versus 9.0 per cent, according to the US census (Ibid). Of these women-owned businesses nearly half (45 per cent) were in the service sector but much of the most recent growth has been concentrated in non-traditional sectors such as construction, transport and agricultural services. These recent trends in the USA suggest that there is considerable scope for expansion of entrepreneurship among women in Hungary.

Koncz (2000) feels that women typically run a different style of business from those owned by men, often creating very small businesses in difficult sectors such as the service sector. They tend to prefer companies that involve limited capital investment and low risk, such as property leasing and retail trade, other than shopkeeping. They are also active in catering where family resources can be used. Where women do operate as shopkeepers, they tend to set up in low risk areas that offer the potential for growth, suggesting that a safe livelihood is more important to them than high profits. According to a 1998 survey of entrepreneurs by the Hungarian Small Enterprises Society, women are more cautious than men about making investments and about increasing the size of their firm (Koncz, 2000). There was some evidence in our study to support these gender differences (see Chapter 5).

According to a study by Gere in 1996 on women entrepreneurs, companies set up by women are more successful than those of men in combining professional life and family (Gere, 1996). Most are in the service sector and seek to develop local potential. They tend to be labour-intensive. Gere (1996) also found that the educational level of female entrepreneurs is higher than that of the average earner. Most were employed in the state sector before changing their employment status, and the majority made the change in the belief, often unrealized, that entrepreneurship would be their most important defence against unemployment. More than half of women entrepreneurs started their business because of the threat of unemployment (Ibid).

The Gere (1996) study also records that 82 per cent of respondents said their lives changed completely after they started their own business. Almost half, however, said that the division of labour within the household did not change and 24 per cent of respondents continued to do the household chores alone but 36 per cent did get more help (Gere, 1996; Hisrich and Fülöp, 1997). Table 1.1 appears to support this trend suggesting that the sharing of reproductive tasks has increased in the last two decades.

However, women still work longer hours than men when productive and reproductive roles are combined and this gender proportion has changed little since 1963 (Table 4.3).

Table 4.3   Total annual working hours by gender in Hungary, 1963–1993 in millions of hours

Year	Women's workhours	Women's % of total hours	Men's workhours	Men's % of total hours	Total hours worked
1963	205	53.7	177	46.3	382
1977	197	53.7	170	46.3	367
1986	186	53.0	165	47.0	351
1993	162	53.5	141	46.5	303

Source: Falussy, Béla and György Vukovich (1996), 'On the balance of time 1963–1993', *Tarsadalmi Riport (Social Report)*, Budapest.

Conclusion

Our working definition of entrepreneurial activity has been to include all those officially recognized and licensed as entrepreneurs. This includes sole proprietors of small businesses as well as working owners of 'corporations without legal entity' (World Bank, 1999, p. 106) consultants and home workers. These are all included in international statistics as self-employed (Gábor, 1997, p. 159). Fieldwork also revealed that some successful small business people are working without official licences, while some register as official entrepreneurs on behalf of other family members. In both of these cases the underlying reason is to avoid taxes. In Hungary corporate taxes are much lower than personal taxes and the 1999 World Bank Report (1999, p. 106) noted that 'almost 70 per cent of the registered enterprises are unincorporated partnerships commonly used to reduce taxes'. The laws relating to taxes and official self-employed status are constantly changing and even in rural areas people are very aware of the implications of these changes: for example, during the field surveys in 1998, we were told that it was no longer necessary for someone providing a few rooms for tourists to buy an entrepreneur's licence. This uncertainty in the context of the transition, while individual economic activity and state regulations are in a continual state of adjustment, undermines the reliability of information on self-employment.

Chapter 5

Gender at the Borders: Entrepreneurs and Social Capital

We surveyed 27 villages which had a total of 1165 officially registered entrepreneurs in 1998. In western Hungary of the 829 registered entrepreneurs in the 17 villages visited, 33.8 per cent were women while in the ten border villages in eastern Hungary, 39.3 per cent of the 336 entrepreneurs were women.

Only two villages, in the west, had more than 3000 people while none of the villages studied in the eastern border areas were that big. Most villages were much smaller, with two in the west and one in the east having fewer than 500 people. The smallest village, Ujszalonta in the east, had only 126 people. By 2003 three villages in the west had fewer than 500 inhabitants. The average population size of the surveyed villages in both eastern and western border areas was similar with 1996 populations of 1578 in the west and 1512 in the east. Some villages had declined in size since 1992 with only two of the ten studied in the east, both located at major international border crossings (see Figure 2.1), having grown. In the west only five of the 17 villages surveyed had decreased in size between 1992 and 1996. Village population decreases were mainly due to low birthrates. Births were exceeded by deaths in all but the smallest village in the east in 1996 for a total of 167 births and 276 deaths. In the west a similar pattern could be seen with only one of the 17 villages, Hegykő on the edge of the National Park, with a rapidly growing tourist trade, having more births than deaths. The total of births in the Győr-Moson-Sopron border villages in 1996 was 241 and of deaths 360. Thus crude birth and death rates were higher in the east than in the west in 1996.

Foreign visitors and foreign guest nights in 1996 were concentrated in five villages in the western border area, while none were reported in any village in the east. In 1996 Hegyeshalom reported 8644 foreign guest nights, Nagycenk 4445, Hegykő 4344, Fertőrákos 1226, and Kópháza 552. Tourism in the west has increased very rapidly since 1996 and now most western border villages offer considerable opportunities for entrepreneurs working in tourism.

In both border areas outmigration from villages, over the four years 1992 to 1996, was greater than immigration, though the difference was least in the west (964 versus 947) whilst in the east the figures were 733 versus 673. Inmigration exceeded outmigration in seven (41 per cent) of the villages in the west and three villages in the east (30 per cent) and the difference between in and outmigration was greatest in the west illustrating the greater attractiveness of this border region. Since the total population of the eastern villages was only just over half (57 per cent) of the population of the western study area villages, it is clear that outmigration plays a greater role in rural border areas of the east than the west. The largest increase (20 migrants) through migration between 1992 and 1996 in the east occurred in Lökösháza, a major rail crossing, while the positive migration balance was highest at 37 in Nagycenk, in the west. The combination of a natural decrease through higher death rates than birthrates and high outmigration is leading to a population decline in these border areas.

History of Entrepreneurial Activity

We were able to ascertain the date of establishment of a business for only about half the entrepreneurs on the village lists. For the eastern border area we had the date of establishment for 71 per cent of the still existing businesses but for the west we only had a date for 40 per cent. Of the seven entrepreneurs who had founded their firms in the 1970s, and who were still in business in 1998, all were men, with two working as hairdressers and most of the others having house building skills. Those contemporary entrepreneurs who founded their businesses in the 1980s are very different. In the west 9 per cent of those still operating had been founded in the 1980s while only 3 per cent of businesses in the east were that old. In the western study area, there was a gender balance in the businesses founded in the 1980s, with the women entrepreneurs mostly running shops or offering hairdressing, dressmaking, or transport skills. In the east three of the current businesses were started by women in the 1980s. One man set up as a tailor in 1980 but it is only after the legal changes of 1982 that women appear in the official record. This gender differentiation supports Róna-Tas' (1997) analysis. In the eastern border region one man, whose father had been an entrepreneur, set up a business making whitewash in 1987 when he lost his state job. His wife had started a business in the previous year mixing animal feed because the state cafeteria where she worked as a cook closed. She had trained as a cook but her husband persuaded her that mixing animal feed was only different in scale from mixing a cake! These two are now among the most successful people in the village, working long hours and ploughing back their profits into the business in order to expand. Their efforts to build up the business are encouraged by their children who are already involved in it. Thus they are more like capitalist entrepreneurs than socialist workers in the second economy, despite being forced into this role before the transition, rather than choosing to become self-employed. The overwhelming dominance of post-communism start-ups in these rural areas suggests that earlier activities in the private sector are of only minor importance in understanding contemporary self-employment.

In both areas there were two peaks of establishment of businesses for those businesses operating in 1998: 1997 when 15.3 per cent of businesses in the west and 15.5 per cent in the east were founded; and 1992 with 13.5 per cent of businesses in the west and 1994 with 16 per cent in the east. This latter difference suggests that self-employment was slower to get started in the east after 1989. In the west seven businesses had survived from the 1970s, while only one, a hairdressing business, founded in an eastern village by a man in 1972, was still operating in 1998. These figures suggest that even before the system change the environment for private businesses was better in the west than in the east. It may also indicate that, in the east, firms have a lower survival rate, though our data does not allow us to prove this.

The Sample Survey

The lists of entrepreneurs provided by the mayors gave information on the type of work done by each entrepreneur, in addition to the date when some individual businesses were founded. In the eastern border region 11.4 per cent of women and 28.4 per cent of men were working in agriculture or forestry. In the west, however, there were fewer entrepreneurs in this primary economic sector: 5.1 per cent of women and 14.9 per cent of men. In the secondary sector of manufacturing, in which we included construction and craft work, there were 9.4 per cent of women in the west and 36.0 per cent of men. In the east, unlike the west, there is no tradition of industrial work in the villages and only 5.3 per cent of women and 13.7 per cent of men were running businesses involving manufacturing or construction. In both areas, the service sector was the dominant one involving more than half of the self-employed, particularly women entrepreneurs. The service sector involved 85.5 per cent of women and 49.1 per cent of men in the west and 83.3 per cent of women entrepreneurs and 57.9 per cent of men in the east.

We interviewed 30 per cent of the entrepreneurs listed in border villages in the counties of Békés and Győr-Moson-Sopron in the summer of 1998. The sample was stratified by gender and by the three major sectors of the economy. In the east 101 entrepreneurs, 48 men and 53 women, in ten border villages, were interviewed. In the west 250 entrepreneurs in 17 villages, made up of 125 women and 125 men were included in the survey. Because of some incomplete questionnaires we have only been able to include 248 of the western surveyed entrepreneurs in some of our statistical analysis.

We also held focus groups and in-depth interviews with entrepreneurs in the two areas. This qualitative information has been used to supplement the quantitative data from the questionnaire survey. The qualitative research took place over a longer period from 1996 to 2003.

Age of Entrepreneurs

The age range of entrepreneurs in 1998 was from 19 to 72 years and both these individuals were men working in an eastern village. Most entrepreneurs were in their forties in both east and west study areas, with an average age of 42 for men and 41 for women in the west and 40 for both in the east. The concentration of self-employment in this age group was more marked in the west than in the east (Table 5.1). More young women than men were entrepreneurs probably because it was more difficult for the women than for the men to look for jobs outside the region (Table 5.1). Young women may have also chosen to be self-employed at this time in their lives when they were coping with small children. Minniti (2003), in her global survey, also notes this tendency for young women to engage in entrepreneurship during the age period when they are most likely to have young children. However, she also found that men were also more likely to be self-employed at a young age which was not true in our study (Table 5.1). We also found cases where a young woman on maternity or childcare leave had taken out the entrepreneur's licence on behalf of her husband, or brother, as she was taxed less heavily. One young woman in an eastern village was the entrepreneur employing her parents in the village shop, as by becoming self-employed she got a

Table 5.1   Age groups of entrepreneurs in eastern and western border villages in 1998

	West		East	
Age Group	Women (N=125)	Men (N=125)	Women (N=53)	Men (N=48)
	Percentage		Percentage	
Under 30 years	17.6	9.6	18.9	10.4
30–39 years	21.6	20.0	28.3	33.3
40–49 years	41.6	43.2	32.1	35.4
50–59 years	12.0	21.6	11.3	10.4
60 years and over	7.2	5.6	9.4	10.4
Totals	100.0	100.0	100.0	99.9*

*Totals do not always equal 100% because of rounding.

Source: Fieldwork.

tax break because she had never had a job. She had trained as a hairdresser but there were already several in her village so working with her parents, in a shop they rented from the village administration, was the most efficient financial solution for the family.

In the western villages there were fewer women than men in their fifties self-employed and fewer elderly entrepreneurs. Many people in their fifties were just working in order to be able to keep paying into their pensions. Until the late 1990s women were able to retire at 55 so many were already on pensions. In the interviews we were often told that at 50 they were too old to start something new. This negative attitude to age was most common in eastern Hungary. Some of the men in this age group were on disability pensions which they were supplementing by working as entrepreneurs. Most of the self-employed retirees aged over 60 in both areas were involved in the mainly subsistence cultivation of a piece of land gained through the privatization process, or had taken out the entrepreneur's licence on behalf of a younger family member who had a business, as the tax was less if the official entrepreneur was a pensioner. Clearly, the existence of a licence in an individual's name did not necessarily mean that that person was actively self-employed but often reflected a family decision in order to reduce taxes.

Size of Firm

Over two-thirds of firms operated by rural entrepreneurs in our survey did not have any employees in 1998 (Table 5.2). In our western study area 68.4 per cent of interviewees were in this position as were 67.3 per cent in the east. The largest firm in the east, owned by a woman, had 14 employees. In the west two firms had ten employees, two had 14 and one had 30 employees. So entrepreneurship was providing some local jobs but only to a limited extent. Most of the self-employed did not want to employ non-family members because the competition for jobs after the transition has led to lack of trust between villagers.

Table 5.2   Number of employees per firm in east and west border villages of Hungary

Number of employees	West (%) N=250	East (%) N=101
	Number of firms	
0	68.4	67.3
1	12.4	17.8
2	9.6	6.9
3	4.4	2.0
4 or more	5.2	5.9
Totals	100.0	99.9*

* Does not sum to 100 because of rounding.

Source: Fieldwork, 1998.

Size of Family

In the eastern villages over half the families had two children while in the west the most common family size was only one child (Table 5.3). However, slightly more families in the east than in the west had three or more children. This difference may reflect the greater level of mobility of women and their professional training in the west and the desire of women in the east to stay home and claim a childcare allowance for three children as an alternative to looking for an elusive job. The number of families without children was just less than one-eighth of the population of women in both study areas. These regional differences in family size do not appear to have limited women's interest in self-employment and the larger proportion of women entrepreneurs found in eastern border villages may indicate that self-employment is seen by mothers as a way of supporting children.

Education Levels of Entrepreneurs

On the whole education levels were higher in the west than in the east, although there was a greater proportion of entrepreneurs with a college education in the eastern border area (Table 5.4). Most rural entrepreneurs had a technical education which had trained them for work on the cooperative. In the general population fewer people completed their education with such an apprenticeship (Table 5.4). This was especially true of men and their skills in car mechanics, and construction had provided the basis for setting up small businesses. In the west we found an unusually low percentage of

Table 5.3   Number of children in families of women entrepreneurs in eastern and western border regions of Hungary, 1998

	West (%) N=53	East (%) N=125
Number of children	Percentage	
None	11.3	12.0
1 child	37.7	23.2
2 children	34.0	52.8
3 children	15.0	10.4
More than 3 children	2.0	1.6
Total	100.0	100.0

Source: Fieldwork, 1998–99.

Table 5.4   Highest qualification of active earners in Hungary (1997) and of rural entrepreneurs in border regions (1998), percentages

	Active earners Hungary		Entrepreneurs Győr		Békés	
Education level	Women	Men	Women	Men	Women	Men
Less than primary	0.9	1.0	0.0	0.8	0.0	0.0
Completed primary	24.2	19.0	8.8	5.7	14.6	20.8
Completed 3-yr apprenticeship	18.2	38.6	32.8	47.2	31.3	18.9
Completed secondary school	40.3	27.3	48.0	42.2	41.6	49.0
Completed tertiary level	16.4	14.1	10.4	4.9	12.5	11.3
Totals	100.0	100.0	100.0	100.0	100.0	100.0

Sources: Lazreg, Marnia (1999), *Making the Transition Work for Women in Europe and Central Asia*, Discussion paper No. 411, Washington DC: The World Bank for the Hungary data and fieldwork in 1998 for the entrepreneurs.

men with tertiary education probably reflecting the different opportunities available for women and men in that area and the importance of female-dominated professions such as accountancy and dentistry.

Types of Entrepreneurial Activity

Table 5.5 shows the breakdown of firms found in the survey by major economic sectors. Agriculture remains more important for men than for women in both study areas but is the basis of more firms in the east than in the west. In the west there are also some men working in forestry in the protected areas along the border. In the east hunting is still an occupation. Secondary sector activities are also more important for men than for women in both areas. In the west there is a long tradition of metal working and this is continuing in small firms. Craft work involves the making of wooden ornaments and china painting for tourists, making of brooms and also dressmaking which occupies several women. Many men are also involved in construction and building. There were eight dressmakers in our study in the west but only two in the east, reflecting different levels of demand. Other small-scale manufacturing in the west includes spectacle making and the production of medical instruments.

Table 5.5   Types of entrepreneurial activity in rural east and west Hungary, by gender

Activity Type	West		East	
	Women %	Men %	Women %	Men %
Agriculture/Forestry/Fishing/Hunting	5.1	14.9	11.9	28.7
Secondary				
Manufacturing/Craftwork	9.4	37.9	5.9	12.9
Tertiary				
Services	85.5	47.2	82.2	58.4
1. Retail sales: goods, food, etc.	50.4	24.8	64.4	38.6
2. Personal services	18.8	0.0	9.9	1.0
3. Professional services	13.3	9.5	5.9	10.9
4. Transport	0.4	11.3	1.0	5.9
5. Tourism	2.6	0.8	1.0	0.0
6. Security	–	0.8	0.0	2.0

Source: Fieldwork, 1998.

Most women entrepreneurs ran small general stores, restaurants or ice-cream parlours while men operated bars and restaurants, hardware or farm supply outlets or worked as carpenters or electricians. Women were also involved in personal services, especially hairdressing, but these opportunities were more common in the west where Austrian tourists and second-home owners provided high-paying customers. A few men worked in security services as guards or watchmen while some were locksmiths. One woman in the west owned a horse-drawn carriage which she used for tourists.

Tourism has also created a demand for handmade souvenirs, and small hotels and restaurants, and has grown very rapidly since 1998 in the western border area, encouraged by the Hungarian accession to the EU. In Békés county one village joined the national rural tourism association in the mid-1990s but failed to attract many tourists, although German hunting parties do occasionally visit. In Győr-Moson-Sopron the mayors of the border villages met in June 1998 to discuss ways of advertising in the western European press to attract visitors, emphasizing especially their local horsemanship. By reaching out across the border to neighbouring countries, western border villages of Hungary have succeeded in building a substantial tourism industry outside the traditional tourist areas of Hungary which focused on Budapest and Lake Balaton. Tourists in this western border region come not only from Austria and Germany but also from the Netherlands, whose cyclists enjoy the many new cycle paths in this flat and picturesque part of the country. German is widely spoken in this part of Hungary and many signs for tourists are in German, thus making it easier for German speaking tourists to visit (Plate 5.1).

In both areas there was a surprising range of professional jobs in the villages, from customs officials and insurance agents to business advisors, lawyers, doctors, dentists and veterinarians. Professional jobs also varied by location and gender: 13.3 per cent of women in the west but only 9.5 per cent of men interviewed had professional jobs, while the gender balance was reversed in the east with only 5.9 per cent of women but 10.9 per cent of men being in this category. Dentistry and accountancy are predominantly female occupations while veterinarians are usually men. In western Hungary, the cheap and excellent dental care attracts people from nearby, mainly German-speaking countries of Western Europe. Some of the biggest new houses in the villages of this region are owned by women dentists. In addition to the women dentists in the west, other women professionals such as accountants, lawyers, surveyors, estate agents and tax advisors found they could make a reasonable living in the village without having to commute to the city. This reflects the level of business activity in this western border area. Professional men in the east were usually veterinarians because of the continuing importance of farming and especially livestock. There was also an agricultural advisor working as an entrepreneur in the east. Doctors did not like working in some of the small eastern villages because of poor living conditions and in one village we met a Middle Eastern immigrant working as a doctor. He was able to get this position because of lack of competition for the village post.

The service sector included over four-fifths of women entrepreneurs surveyed in the study areas. In the east 66 per cent of self-employed women operated small general stores, and restaurants. In the west 50.4 per cent of women entrepreneurs ran retail shops, including general stores and food and clothing stores, ice-cream parlours, wine bars, and restaurants. One successful woman entrepreneur in a village in the west had three clothing stores for ladies, men and children. Her husband, who was on a disability pension, ran the men's clothing store, her daughter ran the baby and children's store while the entrepreneur herself ran the ladies' clothing store with the help of an employee and did all the purchasing. Her youngest daughter was studying business at university so that she would be able to take over the bookkeeping for the firm. In the west the clothing stores included such specialist shops as one renting out wedding clothes and one offering dress design. In the east a high proportion of clothes shops dealt only in second-hand clothes mostly imported from Western Europe, reflecting the much lower purchasing power of villagers in the east. There were nine such shops among the ten villages in the east but only three altogether in the 17 villages

Plate 5.1 Foreign tourists travelling by bicycle and car in western border region. Note multilingual name on restaurant

studied in the west. Other services offered were television and radio repair (three firms in the west but only one in the east) and computer training and sales with all three firms located in the west. The western area also had specialist services such as a photographer and a travel agency. Most villages in the study areas had a specialist flower shop run by a woman, indicating the importance of flowers as gifts in the Hungarian culture.

Personal services such as hairdressing, cosmetology, beauty treatments, manicure, massage and sunbeds were available in the western villages and almost entirely offered by women. Such services were only one-third as common in the east and only included hairdressing and cosmetics because of lower levels of disposable income (Table 5.5) but every village had at least one hairdresser. In one western village there were even two masseuses, while in the east few villages had more than a hairdresser and possibly a cosmetician. This widespread interest in the beautification of the female body is very typical of the former Central and Eastern Europe.

Transport was done almost entirely by men, especially in the west where freight and passenger services were in demand (Table 5.5). The western study area had a total of 63 firms run by men involved in transport and car repair and car detailing while only 14 firms (12 run by men and two by women) were in this field in the east. In the west the car repair and detailing firms benefited from Austrian customers who found these services much cheaper in Hungary.

Tourism was mainly a female occupation and was more important in the west near the source of rich western tourists. Women provided bed and breakfast, with nine hotels in the west and none in the east. There were six souvenir shops in the west but only three in the east. One woman ran a riding school and both men and women ran the two horse carriage operations. Men in this field drove tourist horses and carriages, rented out bicycles or acted as guides in the National Park. One woman in the east who owned several shops in three villages had also provided accommodation and food for German hunting parties. Security, related to the frontier and to the needs of tourists provided opportunities for a few men to become self-employed (Table 5.5). Border differences are reflected in the presence of both men and women customs agents in the west but none in villages in the east.

Table 5.6 provides a schematic overview of the main differences between the two study areas in terms of opportunities for self-employment and the kinds of employment undertaken. The main differences are related to the openness of the border and differences in the number of crossing points (Figure 3.1). The border also influences access to nearby urban centres as both markets and sources of employment. The western location of border villages brings wealth to this area while in the east isolation and lack of cross-border trade makes for a much lower general level of disposable income. This is reflected in the importance of second-hand clothing stores and the number of itinerant traders who have to visit several villages in order to obtain a large enough market, in the east. The lack of purchasing power in the eastern villages can also be seen in the more limited range of leisure activities and food and drink outlets than in the west. Above all, the rural border area in the east is still dominated by agriculture whilst this is much less important in the west. Consequently, the eastern study area with only ten villages had 21 feed merchants compared to only 12 in the 17 villages in the west.

Table 5.6   Regional differences in entrepreneurial activity

Western Border Zone	Eastern Border Zone
Open Border	*Border with Restricted Access*
Local market includes Austria.	Little legal cross-border trade.
Employment created by border crossings.	Few crossings so limited related jobs.
Tourism: dental, second homes, etc.	No tourism: occasional hunters.
Capital from jobs in Austria.	Illegal Romanian workers.
Access to regional urban centres	*Isolated from urban centres*
Importance of private transport services such as trucks, buses and taxis.	Limited access to higher order centres because of border, so only some 'key villages' with petrol stations.
Relatively wealthy villagers	*Poor village population*
Variety of specialist shops.	Mainly general shops.
Second-hand clothing stores make up only 10% of all clothing stores.	Second-hand clothing stores make up over half of all clothing stores.
Hairdressers, cosmeticians, masseuses and manicurists widely available.	Only hairdressers (one per village) and cosmeticians found.
Range of leisure activity outlets: video and bicycle rental, slot machines, sports and computer shops.	Only video rental.
Variety of food and drink outlets: pizzeria, snack bars, cake/tea shops, ice-cream shops, wine bars, pubs, etc.	Off-licences make up 70% of sources of alcohol. More snack bars but only one wine bar.
No itinerant traders.	Many itinerant traders.
Metal workers in every village.	No metalworking.
62% of professionals are dentists, lawyers, accountants, insurance agents, and tax and investment advisors.	No dentists. 29% of professionals are business-related: only accountants, tax advisors and insurance agents.
Declining importance of farming	*Agriculture still dominant*
No veterinarians in villages.	Veterinarians in almost every village (29% of all professionals).
Twice as many beauty shops as agriculture-related ones.	Twice as many farm-related shops as beauty parlours.

Source: Fieldwork, 1998–99.

Reasons for Becoming Self-employed

It is often suggested that people become entrepreneurs because they are forced to do so by the loss of a job or because their employer made them become a private contractor (Table 5.7). In Békés county border villages 41.6 per cent of entrepreneurs, equally divided between men and women, said that they were forced entrepreneurs. In Győr in the west, in the border villages, 21.6 per cent of men but 30.4 per cent of women interviewed said that they were forced entrepreneurs. In some cases our respondents told us that they had set up a business because their spouse became unemployed or was forced to take early retirement because of disability. In other cases young people were forced to become self-employed when they finished their training in technical

Table 5.7   Reasons for becoming an entrepreneur in rural Hungary

A.   Household survival

Characteristics:
1.   Need income despite fear of the risks of self-employment
2.   Forced response to unemployment because of lack of alternatives
3.   Often combined with subsistence agriculture
4.   Part-time to supplement pension or income from low-paying full-time job such as teaching.

B.   Response to fiscal policies

Characteristics:
1.   Division of farm ownership among family members reduces taxes by reducing individual parcel size and value of output
2.   Young person who has never had a job is granted a tax holiday
3.   If husband has a disability pension, the wife registers as the entrepreneur to protect the pension
4.   If the husband is employed, the wife registers as the entrepreneur to protect his job rights
5.   Women on maternity or childcare leave can operate as an entrepreneur without losing benefits and can work flexible hours
6.   Setting up a business in the home allows the entrepreneur to write off the cost of heating and lighting the house as a business expense
7.   Set up business in order to maximize pension on reaching retirement age.

C.   Desire for self-employment

Characteristics:
1.   Family history of business and self-employment, and family source of capital
2.   Wider than average experience outside the region
3.   Choice of occupation based on work done under socialism and/or training
4.   Sees profit and opportunity rather than risk in self-employment
5.   Wants to establish business for children's future.

school as no jobs were available for them in the villages. In both areas teachers said that they had to take on part-time work in order to earn enough to live while several retired people were supplementing their meagre pensions. Some who had trained as teachers found it more profitable to work as an entrepreneur, such as the daughter of the boutique owner in the west who was running the children's store for her mother despite having trained as a kindergarten teacher.

Surprisingly, despite this high level of negative reasons for choosing to become an entrepreneur, 91 per cent of men and 86 per cent of women in the western study area and 91 per cent of men and 87 per cent of women in the east said that they preferred to be self-employed rather than work for someone else. This overwhelming consistency of support for entrepreneurial independence in two regions where the local economy offers very different success rates and opportunities is remarkable. It may well indicate a new attitude to self-employment among the wider population despite the very different mindsets of people in the two areas.

Other reasons for choosing to start a small business were opportunity, such as being able to get land when the village cooperative was restructured, a family history of self-employment, a childhood dream, and the desire for independence and self-improvement. Others did it for the tax benefits. There were also lifestyle reasons such as to give up commuting, to stay at home with children or to allow the wife to stay at home with the children, or alternatively because they were bored at home.

Future Plans

In further support of this positive view are the responses of the interviewees to a question about expanding their businesses (Table 5.8). Once again regional differences were minimal with 29.7 per cent of village entrepreneurs in the east and 31.6 per cent in the west planning to expand. However, such expansion plans differed by gender with more men than women wishing to expand, as noted by Gere (1996) (Table 5.7). Such expansion usually involved opening another outlet in the same or another village, increasing the range of goods or services offered, buying more modern machinery to upgrade production or taking on more staff. Problems specific to women entrepreneurs were pregnancy or increased household responsibilities for elderly family members or children. In the west, where several people ran small businesses on a part-time basis while continuing to work in the main job, lack of time was often mentioned, especially by women, but it was apparently much less of a problem in the east (Table 5.8). There was little gender difference in the wish to stop or give up a business but women were more likely than men to say that they did not plan to change anything (Table 5.8).

The reasons for declining to change were varied and regional differences were marked. In the west lack of space was seen as a problem, especially by women, but lack of capital was a bigger problem, particularly for men (Table 5.7). Insufficient demand was also seen as a problem in many villages especially by women (Table 5.7). Growing competition was limiting expansion among both men and women in the western study area but was not mentioned as a problem by entrepreneurs in the eastern area. On the other hand, increased regulations were seen as more of a problem in the eastern region, especially by men. Age and health were seen as a problem by more people, especially women, in the eastern study area and 15 per cent of the women

Table 5.8   Future plans of rural entrepreneurs in Hungary in 1998

Plans	Eastern Villages		Western Villages		Both Areas	
	Female (N=53) %	Male (N=48) %	Female (N=125) %	Male (N=125) %	Female (N=178) %	Male (N=173) %
Expand	22.6	33.3	24.0	32.0	23.6	32.4
Stop	17.0	14.6	5.6	6.4	8.4	8.7
No change	62.3	54.2	72.0	61.6	69.1	59.5
Reasons for no change:						
Low demand	20.8	14.6	19.2	16.0	19.7	15.6
No space	0.0	0.0	5.6	1.6	3.9	1.2
No capital	11.3	14.6	17.6	12.8	12.4	17.9
Competition	0.0	0.0	5.6	4.0	3.9	2.9
No time	0.0	2.1	16.8	5.6	11.8	4.6
Age/health	11.3	6.3	5.6	8.0	7.3	7.5
Regulations	3.8	4.2	1.6	3.2	2.3	3.5
No wish for employees	0.0	0.0	4.8	5.6	3.3	4.1
Enough	15.1	8.3	4.0	4.0	7.3	5.2
Children uninterested	0.0	0.0	3.2	0.8	1.1	0.6
Needs new skill	0.0	0.0	0.8	0.8	0.6	0.6

Source: Fieldwork, 1989–99.

there felt that their current work was enough. Only in the west was the need for a new skill seen as a problem and several entrepreneurs were already retraining or going back to university for further education. It was also mainly in the west that business people, especially women, based their decision about the future of their firm on whether their children were interested in becoming involved (Table 5.8).

Overall, many of these regional differences in future perspectives reflect differences in both success and opportunities for business in east and west border regions. The short-term results of these different perspectives on the future can be seen in the growth of the number of sole proprietors per 1000 inhabitants between 1996 and 2003 (Figure 6.1). Growth was greater in Győr-Moson-Sopron county than in Békés county. In the west there were decreases in only three of the 17 border villages studied while in the east half the villages had decreases.

Social Capital

Social capital has been defined as 'those characteristics of social structure or social relations that facilitate collaborative action and, as a result, enhance economic performance' (Johnston et al., 2000, p. 746). The term was popularized by Robert Putnam in his study of the economies of the northeastern regions of Italy. He saw social capital as the central explanatory variable of the economic success of these regions relative to the rest of the country. 'What is crucial about these small-firm industrial districts, conclude most observers, is mutual trust, social cooperation, and a well-developed sense of civic duty – in short the hallmarks of the civic community' (Putnam, 1993, p. 161). Putnam's work has been critiqued for its silence on the reasons for regional differentiation in supplies of social capital and for being unrealistically devoid of the potential existence of conflict and contestation (Johnston et al., 2000). Portes (1998), looking especially at the situation in developing countries, saw social capital more in terms of individual contacts and networks. Burt (1998) looked at gender and social capital and noted that although social capital predicts that returns to intelligence, education and seniority depend on a person's location in the social structure of the market these returns differ for women.

In Eastern Europe 'the collapse of communism has resulted in the rapid erosion of virtually all social support systems, and has bred mass insecurity among the people of this region as they have watched their savings and accumulated assets dwindle and disappear' (Narayan et al., 2000, p. 66). As a result of the rupture of the networks developed under socialism, social capital is a new and fragile commodity (Laschewski and Siebert, 2003). This is especially true in small villages where often the only employer had been the agricultural cooperative. The break-up of these cooperatives left rural areas with high levels of unemployment (Kovács, 1996). In many places fellow workers on the cooperative had been considered a second family, especially by women who had often worked in groups rather than individually as the men did (Iganski, 1999). The new competitiveness for jobs created suspicion and lack of trust between neighbours and former fellow workers. Consequently, a post-transition civil society has been slow to develop in these villages. However, rural areas have certain advantages for entrepreneurial activities. The push factors of unemployment and lack of alternative opportunities are often more marked than in urban areas and the positive element of family networks is stronger.

In a study of inclinations towards entrepreneurship based on surveys of 3000 in 1988 and 1000 persons in 1990, Lengyel and Tóth (1994, p. 30) came to the conclusion that the desire to become an entrepreneur in Hungary depends 'more on social resources and job satisfaction than on occupation and level of education', contrary to their original hypothesis. Their concept of social resources is similar to that of social capital, as developed by Portes et al. (1998) but was measured in terms of the number of friends reported by individuals. Social capital in border areas needs to take into account interaction across borders. In our 1998 survey of 349 actual entrepreneurs in border villages, the focus was on linkages to local power centres and the existence of cross-border linkages as well as the potential for such linkages. Thus the emphasis was not just on social networks but also on the nature and potential of these networks in terms of relationships with those who could offer assistance with building a business and also the language expertise and experience outside the country to enable the border entrepreneur to extend contacts across the frontiers of Hungary.

Linguistic Social Capital

The findings indicate both gender and regional differences. In the western study area 64 per cent of men and 78 per cent of women spoke at least one language in addition to Hungarian. In the east the gender difference among entrepreneurs in terms of language facility was reversed with 68 per cent of men speaking at least one foreign language but only 36 per cent of women speaking anything but Hungarian. These differences reflect the influence of ease of contact across the western border versus the limited contacts between Hungary and Romania. The main language spoken in the west was German (48 per cent of men and 58 per cent of women) while many in the east were Romanian speakers (19 per cent of men and 17 per cent of women) and saw Hungarian as the foreign language. Russian, which under socialism had been the main language taught in schools, was the second most commonly spoken language in the east but only one man in the west admitted to speaking Russian. English has now replaced Russian as the most widely taught language in schools but its introduction to the curriculum is so recent we did not expect it to be used by entrepreneurs. However, curiously, 13 per cent of men in the east but no women entrepreneurs spoke English. Two per cent of men entrepreneurs and 5 per cent of women entrepreneurs spoke English in the west perhaps because of the importance of tourism. In the west many of the languages of neighbouring countries were spoken, such as Serbian, Croatian, Italian, Czech and Polish, but no one language dominated as a third language. Two per cent of both men and women in the west spoke French with none speaking it in the east, despite its similarity to Romanian. Family links explain the importance of German and Romanian languages on the borders. In Győr-Moson-Sopron county one man listed Esperanto as his second language and in Békés one man listed Hebrew and Latin as well as German and English while another spoke five languages in addition to Hungarian.

Regional differences in interest in languages are not just at the individual level but also at community level. In Jánossomorja, the largest village in the western study area, the mayor told us proudly that the local high school, which teaches German and English, had two students in the top six in the county for languages that year (Jánossomorja, 1998, interview). In the eastern study area, because of the high proportion of Romanian-Hungarians in the population, there is a Romanian language high school in Gyula which many of our interviewees had attended. Thus the community focus in the east is on maintaining local languages, Hungarian and Romanian, rather than reaching out to new markets by learning new languages.

Social Networks Based on Travel

Some of these languages were also learned while travelling and working in other countries. Most of this travel had been undertaken since 1990. It was assumed that those who had worked outside Hungary would have brought back not only capital for investment but also ideas, practices and contacts they could use in their businesses. In the west 19.2 per cent of male entrepreneurs and 12 per cent of women entrepreneurs had worked outside Hungary but in the east only two men and two women had worked elsewhere, either in Germany or Romania. One woman interviewed in eastern Hungary had travelled to Italy and Turkey to trade in the 1980s but she said that she was now too busy to travel. Several of the entrepreneurs in the east had undertaken trading of goods between Romania and Hungary in the early 1990s but this had declined as prices on both sides of the border converged. By the late 1990s the main contacts were illegal Romanian workers looking for employment on the Hungarian side of the border and some sale of imported goods by Romanian itinerant traders. Such linkages had little effect on the success of Hungarian businesses although the Romanians who were prepared to do seasonal, temporary work for low wages may have been a marginal benefit for agricultural enterprises. One village shopkeeper said the Romanian workers were useful as customers but also blamed them for increasing crime in the area.

In terms of business contacts outside Hungary, whether in terms of customers, investment, or inputs, regional differences were even more marked. In the west 40 per cent of male entrepreneurs and 50.4 per cent of women entrepreneurs had a business link in another country, while only one man, a migrant from Romania, and one woman had formal transborder links in the east. However, several agricultural entrepreneurs in the eastern study area reported occasional foreign buyers, usually German but also from Croatia and the Czech Republic, for produce such as pumpkin seeds and paprika. In this case proximity to a border was not the reason for these contacts, but rather the importance of horticulture in this region. In the western study area Austrians and Germans were regular customers for services such as dentistry, cosmetology and car repairs because prices in Hungary were lower. However, a boutique owner in Jánossomorja said that her customers were only those who worked in the village as village residents who worked in Austria earned enough to shop in the city of Győr, where prices were higher but there was more choice.

Local Social Capital

In terms of local linkages to power centres in the village and as a measure of the position and role of entrepreneurs in the village, the relationship between entrepreneurs and local government was studied. It has been suggested that successful entrepreneurs are unpopular in villages because of the 'tall poppy' syndrome, and during the survey and interviews this type of situation was reported by some participants. Such entrepreneurs usually depended on trade with several villages rather than just the village in which they were located. But for the majority of entrepreneurs who provided services locally, their success depended on popularity in the village. One measure of this is whether the entrepreneur has been elected to local office or has a family member (spouse, children or parents) who is an elected official. In Győr five men (4 per cent) and one woman (0.8 per cent) were elected representatives on the local council. In Békés county the gender balance was reversed with no male entrepreneurs but three women (5.7 per cent) being elected to the council. One very successful woman entrepreneur in the east reported that she had been nominated for election to the village council by her brother and husband. When she asked them why they had nominated her, they told her that people in the village liked her better than them. She also chaired an important council committee. In the west 5.6 per cent of men and 4.0 per cent of women had a close family member as an elected member of the village council. None of the male entrepreneurs interviewed in the east had such a family member while one woman entrepreneur's husband was on the council. Two women entrepreneurs surveyed in the east were also employed by their local councils.

Local networks were further investigated in terms of whether the entrepreneur thought that they were known for being outspoken in village meetings or otherwise seen as being an opinion-maker. Here the gender balance was the same in both areas.

In the west 4.8 per cent of men and 8.0 per cent of women entrepreneurs, and in the east one man and eight women (2.1 per cent versus 15.1 per cent) considered themselves to be outspoken on village matters. In both cases women may find it easier to speak out than to get elected, but the greater role of women in the civil society of the rural east may indicate a lower proportion of men living and working in these villages, allowing women to assume these leadership roles.

This interpretation is reinforced by the greater importance of business connections to local government in the west and among male entrepreneurs. In Győr 19 per cent of men and 17 per cent of women said that their links to the local government were through their business while in Békés 13 per cent of men and 8 per cent of women reported this link. Although this relationship may in some cases have amounted to no more than getting paperwork done at the local government office, it was noticed that in each village local officials were well acquainted with the most successful entrepreneurs in the community. In the west many of the entrepreneurs had used the networks provided by their former jobs to find both suppliers and customers. In fact some had started as outworkers for their former employers as urban factories found it cheaper to outsource work rather than pay social benefits to employees. Such links to former jobs were not evident in the eastern study area.

Analysis of Social Networks

A Principal Components Analysis (PCA) of the data was carried out for the whole study area, the eastern and western areas separately and men and women entrepreneurs separately. Principal Components Analysis identifies the underlying structure or components of a data set, extracting components in the order in which they explain the greatest proportion of the overall variance in the data. We included in the PCA 15 composite variables related to household structure, size and location of business and aspects of social capital (Table 5.9). For the study as a whole eight components accounted for 69 per cent of the overall variance (Table 5.10). The first component, explaining 10.85 per cent of the total variance, was identified by high positive loadings for extent of foreign connections and language skills, and the number of businesses operated by the interviewee. This suggests that the main difference in the data set is a distinction between the more successful entrepreneurs who were reaching out to customers and sources outside Hungary and those who remained limited to their own community. The second component, explaining 9.61 per cent of the variance, was linked to the age of the entrepreneur, and number of children they had. The third component linked the number of children and the help obtained for childcare, especially in terms of official maternity leave, and the number of enterprises operated and explained 9.10 per cent of the variance. The fourth component, explaining 8.46 per cent of the total variance, appeared to identify families with several linked businesses. These were often agricultural enterprises which were usually farmed as one unit but registered in separate parcels by different family members to minimize taxes. The fifth component (8.23 per cent of variance) was identified as related to businesses run by husband and wife versus help from relatives distinguishing between entrepreneurial nuclear families and those with wider family links. The sixth (7.99 per cent) was linked to business size as measured by location and number of employees, with the small enterprises run out of the home least likely to have several workers. The seventh

Table 5.9   Variables used in Principal Components Analysis

---

1. Age of interviewee.
2. Number of children in household.
3. Childcare-sum of type of state care used: maternity leave, kindergarten, after school care.
4. Unemployment: number of months of unemployment plus number of separate periods of unemployment.
5. Activities of enterprises operated by interviewee: ranked 1–5 so that highest score has the greatest variety/diversification.
6. Location of business: in home, village, neighbouring village, town, several locations. Ranked 1–5.
7. Number of employees.
8. Number of family members with businesses.
9. Nature of spouse's enterprise: 1 = subsistence agriculture with some production for sale; 2 = self-employment; 3 = cooperative; 4 = independent firm; 5 = limited company; 6 = company with shareholders.
10. Nature of enterprises of other relatives: 1 = subsistence agriculture with some production for sale; 2 = self-employment; 3 = cooperative; 4 = independent firm; 5 = limited company; 6 = company with shareholders.
11. Links between family businesses.
12. Degree of foreign connections: 1 = neighbouring country; 2 = Western Europe; 3 = other.
13. Number of foreign visits as tourist or on business times: 1 = neighbouring country; 2 = W. Europe; and 3 = other.
14. Number of languages spoken by interviewee.
15. Sum of interviewee's level of contact with local government: as employee, elected official, outspoken participant in council meetings.

(7.53 per cent) again reflected family dependence with the number of years of unemployment linked to assistance from relatives. The eighth component (7.45 per cent) had its highest loadings on travel abroad and links to local government perhaps identifying those entrepreneurs who were most cosmopolitan and influential in their villages.

For data from Békés County in the east eight components accounted for 71 per cent of the total variation but the identification of the components differed. The first component, explaining 11.89 per cent of the total variance, was linked to childcare and the number of children, the second to links between family businesses, the third was linked negatively to age and spouse's business perhaps recognizing young unmarried entrepreneurs. The fourth component linked aspects of social connections including travel and foreign connections and links with local government. The fifth picked up size and location of enterprise, the sixth had only one high loading variable, that for unemployment. The seventh component had high loadings for help from relatives and number of enterprises while the eighth had a negative link between language facility and links to local government. This last component may indicate

Table 5.10   Ranking of Principal Components extracted by region and gender

Component identity	Ranking				
	W. Hungary	E. Hungary	Women	Men	Overall
Age of ep and number of children	1	1	2	2	2
Childcare and number of children	5	1	3	4	3
Language skills/foreign travel	2	4	1	1	1
Links between family businesses	4	2	4	3	4
Age and relatives' businesses		3			
Foreign connections/business activities	6	3	1	1	1
Foreign connections and # businesses		6	1	1	1
Foreign connections/links with local government		4	5	8	
Size and location of enterprise	3	5	6	5/7	6
Time spent unemployed	8	6	6	6	7
Unemployment/relatives' businesses			6		7
Spouse's business v. relatives' businesses	7			7	5
Language skills v. local government help		8			1
Relatives' businesses/eps' businesses		7			
Links with local government	8	4	5	8	8
Family businesses		7	8		
Relatives' type of business		4	7		

Ep = entrepreneur

Source: Questionnaire data.

that Romanian speaking villagers were less likely to have close links to local government than those whose first language was Hungarian, suggesting the continuing importance of ethnic barriers. It is also noticeable that social networks both with overseas contacts and with local government were much less important than in the west.

In the richer western County of Győr-Moson-Sopron, the eight components explained 70.5 per cent of the total variance. The first component, explaining slightly less than in the east (10.21 per cent of the total variance), was identified by high loadings for age and the number of children, perhaps reflecting a division between women in rural villages on the basis of age and size of family as noted by Kwiecińska-Zdrenka (2001). The second component emphasized the link between foreign travel and language skills. Interestingly, this relationship was not seen in the east where travel to other countries was usually short term and was undertaken for trade, often in the form of illegal smuggling. In the west travel was generally more long term and involved employment in well paying jobs in Austria where knowledge of German was necessary. In addition, several people in the west were sending their children to study in Germany or Austria. The third component for the Győr data reinforces the different aspect of social capital in the west. On this component, the highest loading variables were size and location of enterprise with degree of foreign connections. The fourth component was linked to relative's businesses, the fifth to childcare and children, and the sixth linked the number of enterprises operated with foreign connections, again emphasizing the importance of crossborder links. The seventh component was identified with the spouse's enterprise and negatively linked to help from relatives and to local government. This component may be recognizing nuclear families with a strong entrepreneurial tradition and no need to depend on relatives or local government. The final component had its highest loading for time spent unemployed and was negatively linked to local government contacts suggesting that only the successful had good relations with local government. These last two components seem to indicate less dependence on family and local connections in the west than in the east. Lengyel and Tóth (1994) noted that in the period of the transition family background was important in explaining positive attitudes to entrepreneurship but its role was declining. It would seem from the 1998 survey data that it had declined more in the west than in the east.

Principal Components Analysis by gender revealed marked differences in the importance of social capital in explaining overall variation. Eight components extracted 72.63 per cent of the total variation for women entrepreneurs and 70.71 per cent for men entrepreneurs. For the women in both regions the first and most important component, explaining 10.68 per cent of the variance, was related to language ability, foreign contacts and number and size of businesses operated, suggesting that the most successful women entrepreneurs were those with the most links across the border. The first component for the analysis of men entrepreneurs was similar but also included foreign travel but not number of employees and it explained more of the total variance at 11.54 per cent. This component appears to be identified with self-employed lorry drivers and traders involved in cross-border transport. For both men and women the second component was linked to the entrepreneur's age and number of children. For women, childcare and the number of children loaded on the third component while these variables were associated with the fourth component for men. This difference in ranking underlines the continuing responsibility of women for childcare and its impact on their choice of self-employment.

However, for men and women entrepreneurs in the east the first component, explaining over 13 per cent of the variance, was identified with high loadings for number of children and availability of childcare. In the west the first component for the analysis of men entrepreneurs, explaining 11.86 per cent of the variance, was linked to foreign connections while for women the two most important variables on this first component, explaining slightly less of the total variance at 10.53 per cent, were age and children, differentiating between older women with children and younger women without. The importance of foreign social capital showed up on the second and third components for women while the importance of age and children was relegated to the fourth and fifth components for men. In the east local government contacts (Component 3) were more important for women than foreign contacts (Component 8) while for men the reverse was true with language ability and foreign contacts defining Component 2 and local government links relegated to the last component extracted. In the west social capital derived from links with the local government loaded highest on the sixth component for both men and women.

These gender and spatial differences, as shown by the Principal Components Analysis, seem to underline the permeability of the western border with its nine crossing points with 17 types of crossings versus the east with only eight crossing points (Figure 2.1). They also suggest that cross-border trade and collaboration is more important than local community-based trade in the west while in the east the opposite is true. In the east the number of children and the availability of childcare loaded highly on the first component for both men and women. In the west, however, although children had their highest loading on the first component for women entrepreneurs this variable was relegated to the fourth component for men. Availability of childcare loaded highest on the fourth component for women and the fifth component for men entrepreneurs in the west. These gender differences suggest that for women in the west entrepreneurship is a way of coping with childcare by working from home while for men, children and their needs had little influence on their business activities.

Forced Entrepreneurship

Although in both areas studied, many people said that they were 'forced' entrepreneurs only becoming self-employed when they lost their previous jobs, the unemployment variable had few associations with other variables in the analysis. For male entrepreneurs in the study it was linked with help from relatives on the sixth component, while for women it was linked to the location of the business but also loaded highest on the sixth component. These differences suggest that men turned to the family for help when they lost their jobs while women chose to stay home in the village and set up a business. When broken down by both area and gender further differences occurred with the unemployment variable accounting for more of the overall variation in the east than in the west although the gender difference remained. In the west, help from relatives was associated with other forms of social capital – for men with links to local government and for women with language skills and foreign contacts – while in the east help from relatives for men was associated with age while for women it was associated with linkages among entrepreneurs generally. This regional difference suggests that in the west those with strong family links also made connections both inside the community and outside while in the east it suggests that for men dependence on family support increased with age, perhaps reflecting the large number of middle-aged men on disability pensions in the east. However, 'many men who faced imminent dismissal from their jobs bypassed the unemployment registry, opting instead to qualify for disability benefits' (Kulcsár and Brown, 2000, p. 115). The advantage of disability benefits over unemployment is that those considered disabled receive limited but certain incomes while unemployment benefits are for only a limited period of time. In the early 1990s there was considerable official 'flexibility' in granting disability status and it appears to be more prevalent in eastern Hungary where there were few employment possibilities (Ibid).

Conclusion

One of the main differences between the western and the eastern border zones was that in the west entrepreneurs started their businesses because they saw an opportunity and unmet demand in their local community. They often took on this activity as a second, part-time occupation while maintaining their main job. On the other hand, in the east many women interviewed said that they saw entrepreneurship as a last resort when they were unemployed or husbands were unable to support the family. Yet at the same time they showed pride in their own business acumen and their achievements. Many couples run their businesses jointly although the woman may be registered as the entrepreneur, even though she may also have a full-time job. In both areas entrepreneurs turned first to family for capital and employees. The close social networks of the villages had been based on the agricultural cooperatives and after these were broken up so were social relationships. Yet the paternalism of the old large firms in rural areas remains and the development planners operationalizing the new European Union projects tend to treat local actors as subjects rather than as actors themselves (Laschewski and Siebert, 2003).

Kalantaridis (2004), in his study of entrepreneurship in 12 Western European rural areas, found that need-driven entrepreneurship is most important in hostile socio-economic environments, as was seen in this study comparing east and west Hungary. He also reports that the ability of entrepreneurs to leverage resources from outside the area is crucial to success (Ibid). These findings from the broader study support those of this study of Hungarian rural entrepreneurship. In both border areas of Hungary economic success as an entrepreneur is both a cause and an effect of the individual's level of social capital. The greater prosperity of the west, encouraged and reinforced by active cross-border networks, has led to a faster growth of new social capital than in the east. This regional difference will be expected to increase now that Hungary has acceded to the European Union, with the western border becoming even more open and as human capital related to mobility (Urry, 2005) becomes more important.

Chapter 6

Gender Relations in Entrepreneurial Families

It is often considered that as women earn more they gain in respect and in decision-making within the family. Given that women had been almost as economically active as men under communism, we wondered if entrepreneurial activity would have any effect on gender relations in the family. We used individual life-histories to look at the decision to become an entrepreneur and the impact of this at the household level (Sizoo, 1997; Oughton et al., 2003). In interviews with successful women entrepreneurs in both study areas at the time of the main survey, we found some surprising variations at the household level. When we asked husbands if they would be willing to allow their wives to work if they did not need the money for the family, men in the west were overwhelmingly supportive of their wives: 'Its her choice, she should do it if she enjoys it.' In the eastern border zone the opposite, very patriarchal view was expressed: 'It is her duty to stay home with the children.' In the west we even had one man who had taken childcare leave from his job to look after their sick daughter so that his wife could take a training course which would benefit her business. In follow-up life history interviews in 2000 and 2001 in the east, requests to talk to wives were often turned down by husbands and many women entrepreneurs appeared to have gone out of business.

In Budapest and in the counties along the western and north western borders, currently the most prosperous regions of the country, women's economic activity rates closely approximate men's. By contrast, the counties in the east of the Great Plain and in certain parts of northern Hungary exhibit the most marked gender inequalities in terms of access to paid work and are the economically most backward regions of the country, where female economic activity is at its lowest. Such uneven spatial development and its impact on gendered employment may also add to social inequalities between women and men.

In order to explore the relationship between uneven development and gender relations at the household scale, we conducted an empirical study[1] prior to and, in part, simultaneously with the survey analysed in this book. We compared the survival strategies adopted by rural households in Győr-Moson-Sopron and Békés Counties, our two study areas, in response to the regime change, as well as male and female contributions to such strategies (Timar, 2001; 2002).

Our studies confirmed that three forms of adjustment to increasingly difficult macro-economic conditions have become widespread: firstly substitution of home produced food and services for previously purchased consumer goods and services; secondly

---

[1] A project on household survival strategies sponsored partly by the Hungarian National Research Fund (OTKA No. T020443).

reduction in overall household consumption; and thirdly reliance on state redistribution, the occasional bank loan, or financial help from friends or relatives. Smith (2000) also noted similar findings in his studies.

The selection of household survival/livelihood strategies is heavily influenced by differing spatial possibilities with the regional differences highlighted in our study unmistakably present. The main difference between Győr-Moson-Sopron and Békés Counties in terms of the strategies adopted is that in Békés County employment opportunities, especially opportunities for part-time jobs, are much fewer. The impact of the proximity of the state border also varies. The border is of no special importance in Békés County, except perhaps in Méhkerék, where the illegal employment of workers from Romania by greenhouse owners is relatively common. By contrast, along the western border, most households expressly count on the proximity of the state frontier if they have to respond to changing external circumstances and seek a new source of income. Though less emphasis was laid on the urban/rural dichotomy, it can be identified in the various types of adjustment, especially in areas that are a long way from cities. Interviewees refer to this in connection with lack of job opportunities and the difficulties of commuting. Also, complaints were voiced that reasonably-priced merchandise is harder to come by in rural areas than in cities. We conclude that those living in disadvantaged rural border areas face not only greater economic difficulties, but also their means of adjustment to these conditions are more limited. Regional and rural–urban inequalities produced by the new capitalist uneven spatial development are reproduced at the household scale.

The life stories also revealed that survival strategies do not necessarily mean equal sacrifice on the part of all household members. When they opt for reduction in consumption, for example, many families decide not to purchase electrical appliances that would save labour, especially that done typically by women, in the household. Moreover, sometimes household appliances, already purchased, such as automatic washing machines that use a lot of water, are no longer used. There were a few cases found in the study confirming that the 'economic advantages' thus gained lead to tension in gender relationships, signalling the relativity of the rationality of action and choice (see also Wheelock and Oughton, 1996).

For many families today childcare benefit is the only source of income, and the only possible solution to economic difficulties, especially in eastern Hungary. For people in the lowest income brackets, whose plans do not include sending their children to universities or colleges, having a third child may be a kind of survival strategy. Younger female interviewees included a relatively large number of mothers with three children. These mothers with three or more children are allowed to stay at home and are provided with a modest income, and so avoid unemployment. For their families, low as this regular income may be, it can be an important source of livelihood. Lazreg (2000) points out that childcare benefit was a measure deliberately introduced in the mid-1990s in order to reduce the size of the workforce. This legislation suggests that the state is encouraging a resumption of traditional female roles and perhaps also hoping to increase the fertility rate which nationally is below replacement level at an average of 1.2 children per woman (The Economist, 2005). In our study of entrepreneurs we also found a higher proportion of families with three or more children in the eastern border area (see Table 3.5) where opportunities for jobs are harder to find than in the west. Thus it appears that this survival strategy is not limited to the poorest families.

The study of survival strategies provides baseline information against which to measure the situation of entrepreneurs and the relationship between self-employment, uneven spatial development and unequal gender relations.

Focus Group Results

The extensive quantitative questionnaire survey of entrepreneurs in the two border regions was supplemented with further research based on qualitative methods. The nine preliminary interviews conducted early in the project in 1997, 1998 and 1999, were followed by two focus group discussions and additional in-depth interviews in 2000 and 2001. This book provides an opportunity to give feedback on the results of these surveys to the people involved. The success of the focus groups differed in the two regions, providing further evidence of the existence of regional differences. These focus group programmes, both lasting nearly a whole day, were organized in the spirit of mutual help, with the researchers providing information on issues relevant to entrepreneurs and staging a mini-training on 'how to handle problems and stress' for members of the focus groups. In the south east, seven out of ten invited female entrepreneurs managed to attend the discussion, whereas in the north-west only two came. These two participants explained that the low turnout was not attributable to indifference, but rather to the fact that most entrepreneurs were very busy on that day of the week. 'On Fridays transit traffic [across the Austrian border] is heavy, so entrepreneurs cannot afford to close.'

In the west the main problem indicated by the women present at the focus group, apart from shortage of time, was gender relations within the family. There was a consensus that husbands were resentful of successful women entrepreneurs. In the east the main problem was seen as related to health. Many women had husbands who had retired early on disability pensions. It has been suggested that such a status may sometimes be a way of avoiding the stigma of being unemployed when their jobs were lost. In the cases in our study the disability appeared to be fairly minor. It may have been enough to prevent an individual doing heavy manual work but did not hinder them in lighter work associated with service-oriented self-employment. In many of these situations the woman was the official entrepreneur for tax purposes but the man was doing much of the work. The focus on health appears to be related to the post-transition problem of having to take responsibility for their own health care and to pay supplementary fees to doctors. In the west people seem to have adapted more quickly to a proactive approach to health care as is found in Western Europe. Such differences in attitudes may well reflect the isolation of people in villages on the eastern border with no nearby examples of different lifestyles, as occurs in the west.

Life Histories

During the life history interviews conducted in 2000 and 2001, we found changes that had occurred since the questionnaire surveys were conducted. Our objective was to ask female entrepreneurs and their respective husbands about their lives. We intended to choose women who were successful, engaged in a wide variety of businesses, married, of differing ages, living in different villages and who had, at the time of the questionnaire survey, business development plans. We were especially interested in their family, education, jobs, living conditions, roles in the family, relationship with their respective spouses and children, future plans, opinions on the village in which they lived and on their wider environment and the track record and status of their businesses. Wives and husbands were interviewed separately. In the south eastern border region, as a result of the negative economic trends in the years that had passed since the questionnaire survey, out of the 12 entrepreneurs selected from the questionnaire survey database for interview, seven either had gone out of business or declined our request to participate in the discussions, owing to their disappointment with self-employment, or the fact that they no longer worked in the business that was owned by their husband. In this region there was also one successful female entrepreneur who, in the meantime, had assumed public office, and repeatedly declined to see us, citing her busy life. Thus, four female owners of four businesses and their respective husbands were interviewed in Békés County, while six female owners of six businesses and the husbands of four of them were interviewed in Győr-Moson-Sopron County.[2] Overall, 18 persons were interviewed. Relying also on the findings of the earlier interviews and the focus group discussions, our findings are summarized in the following pages.

The Role of Household Gender Relations in the Emergence of Private Businesses

The stories told by our interviewees about how they became self-employed may, of course, be interpreted as their responses to the challenges of individual life situations. These stories have, however, shared characteristics on the basis of which our interviewees could be put into different groups. Obviously, some of the businesses in rural areas were set up by their female owners as part of their own adjustment to the new socioeconomic situation. Others were not. That four out of the ten businesses whose owners were interviewed in the new millennium, were established at the suggestion of the husband rather than on the initiative of the wife, clearly indicates the penetration of patriarchal gender relations into the local economies of the villages included in our study. However, rural households in many parts of the world are strongly patriarchal or as Heather et al. (2005, p. 93) put it 'the social reproduction of patriarchal subjectivities in rural communities' is so woven into rural discourses that it may have led to the story told to us being more fitted to local patriarchal norms than fully taking into account the intra-household bargaining behind the decision to become self-employed.

Female-owned Businesses Established on the Initiative of the Husband

The interviews in this group included one business operating in the north-western border region of Hungary and another in the south-eastern one, which were set up at the husbands' insistence. Neither is directly related to the husband's business, which in each case was set up well before the change in the political regime, but they have

----

[2] Two women entrepreneurs did not want us to talk to their husbands.

become integrated and are now family businesses with employees and adult children as 'business partners'. Mr A, in the east, followed in the footsteps of his father, who was a bricklayer and carpenter and had obtained a business licence as early as 1968. Not surprisingly, Mr A, who had worked for him, also became a self-employed carpenter. His father retired in 1982 and Mr A had an accident in 1983 which affected his hands and forced him to give up this line of business. He switched to animal husbandry and whitewash production and distribution. One of the most successful businessmen among our interviewees in Békés County, he was considering doubling the number of pigs being reared at the time of the interview, and was already running four feed shops. Mrs A, his second wife, was a trained cook and had found employment in the village where they currently live. She commuted to work daily from her natal village nearby. Responsible for two kitchen assistants from the outset, she worked in a canteen operated by ÁFÉSZ (the State Consumption and Sales Co-operative). It was there she met her husband. One month after they got married, in 1977, she gave up her job in the canteen.

By contrast in the west, Mr B, of a similar age to Mr A, had been a skilled worker in the machinery manufacturing division of a large transport company operating in the Western border region. One of his colleagues decided to set up his own business, and he joined as an employee. A couple of years later, in 1982, as he could not get along with his employer, he decided to start his own business. His wife, who had been trained as a postal worker, was employed at a post office in a small town. At the beginning of the post-socialist era, she gave up her job to join her husband's machinery manufacturing business as an 'unpaid family worker'.[3] Then in 1996 she started her own business.

Both female interviewees admitted that they had decided to change jobs at the urging of their respective husbands. With the family home doubling as business premises, Mrs A helped her husband and father-in-law with the sale of building materials from very early on in their married life. Her husband persuaded her to give up her job in the canteen and stay at home. After his accident, and the closing of the restaurant in 1986 he proposed that they should open an animal feed mixing business with Mrs A as its owner, so that 'she might earn social security service time'. However, an underlying reason for his proposal, he admitted, was to find a way of keeping his wife at home.

> You know that just didn't work . . . , there was also the cooking and the washing to be done. She went to work in the morning, or if there was a wedding [they prepared food for wedding receptions at the canteen], she stayed the whole night there. So, that was not what you'd call a family life, was it? This didn't work with us. And then, housework was left undone . . . Yes, she gave up being a cook.

With an ingrained pattern of gender relations passed down from his parents, recalling his mother's life, he voiced an even more unambiguous opinion about the duties of a wife and about women's pay, which was lower than men's even in the socialist era.

---

[3] This is an officially registered status for those who work in a business but do not necessarily get paid.

My mother came from a peasant family. She worked at the cooperative farm. They made her work for peanuts. Then, she stopped going out to work. She stayed at home to look after us; she was a housewife. She didn't want to quit her job [at the cooperative], but there was no one else to cook for us .... And that just didn't work.

He did not think it had been difficult to persuade his wife to work at home. Mrs A, however, cherished memories of her time as a cook at the canteen. Everybody had been pleased with her performance, and she had liked being able to earn a living, no matter how meagre her pay had been. 'I was sad when I had to leave because I liked it there. And they, too, asked me not to quit, as it was going quite well. But I gave in, and agreed to stay at home.'

Later, when her first child was almost three, she went back to work in the restaurant. It was a different experience, though, the second time as, with a new boss, she no longer liked it as much. Finally, in 1986 the restaurant closed and so she went back to working in the family business, which had expanded considerably in the meantime, and now kept her very busy.

My husband is still thinking about expanding the business. And there's my son, Miklós. He's also hell-bent on it. But I'm past 40 now. My old man is nearly 50, too. [They are only three years apart in age.] That's too much for us . . . I said so. I'm getting tired. He said I wouldn't have to work any more, but it's still very much the case.

In the western border region, Mrs B also liked her job at the post office. She did not mind even doing shift work when she went back to work after her childcare leave for her second child had expired.

But when my husband set up his business, shift work became an increasingly pressing problem. I had to work on Sundays, and at weekends. It was he who had to accommodate to my job, look after the children or find somebody to do it . . . He kept trying to win me over . . . It was difficult for me to say yes. First, I was registered as an unpaid family worker. Then, when our son was eight, he said we should now make up our minds, as the business had been keeping him busy. What was he to do? Should he hire somebody? Or would I stay at home? And then I decided I would.

The reason why Mrs B applied for a business licence in 1989 was that the flat rate tax regime for businesses was introduced then. Thus, she opted for flat rate taxation, while her husband followed the cost deduction system. These tax arrangements helped to ensure the profitability of their respective businesses.

According to Mr B's memory of this period, with two children around, his wife's going out to work was 'a pain in the neck'. So, as he put it,

We talked it over. Well, yes, sort of. He conceded that it was difficult for her, seeing all the work she'd put in for the same employer for sixteen plus years. That made it difficult, there's no denying that. But time has justified our decision. It was the right thing to do.

As he sees things now, his business was already quite stable, so it did not matter that his wife earned somewhat less in her new role than she used to. The main thing for Mr B was that 'things got better as she was always at home'. Mrs B also concluded, on a positive note, that her contribution to their prosperous ironworking businesses, which now also included both their sons, had been preserving the values of hearth and home. The greatest benefit was that they could have lunch and supper together. This had created 'strong family ties between the four of us'. The question remains, though, at what price? She was involved in the manual phase of manufacturing a certain spare part. She, however, found this rather stressful.

> ... and that was my responsibility from morning till night. It was too much, and my hands ached, too. But the worst thing was that I developed some sort of allergy. My hair started to fall out. We didn't know what was going on. Then we found out that I was allergic to the metal.

They have had an employee since then. Even so, she keeps working in the business though her husband is against it.

In the case of the other two women included in the group whose members started a business on the initiative of their respective husbands, we found a stronger link between the regime change and their setting up a business. What they further have in common is that their businesses are run as a supplementary economic activity, and only for 'technical' reasons and in order to 'save costs' were they owned by them rather than by their husbands. Despite these similarities, the female owners' attitude to their businesses and their ability to stand up for their own interests vis-à-vis their respective husbands are rather different.

In the east, Mrs C, at 34 years of age, and with a secondary school certificate, gave up her job at VOLÁN (a transport company), which she liked, in order to move in with her (then) would-be husband, who was divorced. A foreman at the construction division of a machinery manufacturing factory in a small town, Mr C had had previous experience in earning extra income through an economic association operating within his company. As he had not got along with his superiors, he started to work, in the mid-1980s, as a sub-contractor for a former colleague, who had set up a business. He was responsible for a team of bricklayers. Satisfied with the money he was earning at the time, he did not want his wife, who was staying at home with their children, and receiving childcare allowance, to ever go back to work, though he knew that she missed her former colleagues.

> I told her that she should stay at home. I could earn the money that she would receive by putting in a few extra hours. I'd be home just a few hours later. The house would always be clean and tidy and the children properly looked after. There would always be a cooked meal when I got home, with the place kept tidy. No rush. We could even have the weekends to ourselves.

However, around the time of the regime change, construction slackened off, with a drop in the number of orders. Moreover, Mr C felt exhausted and overworked. Thus, they seized an opportunity that arose in the husband's natal village: they moved into a company flat, and Mr C replaced his father as a dyke-keeper. They 'went home', downshifted and settled for safer public employment. Though Mrs C was also given a part-time job as a dyke-keeper, which did not keep her away from their three children and the home, even the two salaries combined were too small. So they had to find some way of earning extra money. First, they raised cattle, then they ran an animal feed shop and sold feed in the neighbouring villages. Although the idea came from

Plate 6.1   Husband and wife entrepreneurs in western border region. They have built a metal workshop attached to their house and have a car and van in the yard

Mr C and the work was related to his 'core business', they decided that the business should be run in the wife's name. In this way, they hoped to avoid any conflict of interest that might have arisen in connection with the husband's part-time job at the local hatchery. Furthermore, they wanted to minimize his taxable income that served as the basis of the assessment of the compulsory child maintenance payment due to his child from his first marriage.

According to Mr C, the business, which had had its ups and downs and had been revamped repeatedly by the time our interview was conducted, was not set up because of economic necessity. They simply seized an excellent opportunity that had arisen; and there were no up-front costs to be paid. By contrast, Mrs C did refer to 'necessity' when 'there was this temptation'. Although she did not like the place, she had never perceived this business as an opportunity to break free from their current life, with the issue of going bigger 'leaving her cold'. 'It is not I who run the business. He has it in him. I have only lent my name.' Even if they sometimes have a row over her husband's new ideas about development and revamping, she is still involved in the business. She helps out with whatever is needed. Furthermore, when they decided to open a second-hand clothes shop, and in order to do so, she needed a new qualification, she went to a trade school. As she put it, though it was difficult to bring herself to do so, she succeeded in persuading herself not to go back to work, reasoning that they would not have more money left because of the travel expenses.

> In addition, 'the children would all be his problem'. Yet, she could not suppress her concern, saying 'what bothers me now is that I'm always at home. And also, I sometimes have the feeling that I'm getting duller by the day'.

Also in the east, for Mrs D, just the opposite is the case. A doctor's receptionist and assistant, she was, as she put it, well-loved by the locals. So much so, that she was considered 'the person of the village' and had been elected to the local council. She was aware that her salary was paltry compared to the income from her husband's business, yet, she thought her work was important. She was proud that she had never had to depend on anybody for a living. This seems to have been all the more important for her, as her mother's example was a lesson that had remained with her for the rest of her life.

> Animal husbandry, which was the most difficult of all things, was always my mother's responsibility. My father always left for work early, so my mother had to tend to the animals. She had to carry this burden so that my father might have a job and thus become eligible for old age pension . . . What she did didn't count towards social security service time . . . My father died prematurely at the age of 56. And I had to witness the dire situation in which my mother found herself. She was eligible for a widow's pension for one year only. Then she was no longer eligible, and there she was with no income to rely on . . . Nothing whatsoever. This gave me pause for thought. And I vowed that I would never get into a situation like that if I could help it.

She stands up to the efforts made by her husband, the owner of a farming business, to keep her at home. 'She'd easily make as much as she makes at work here at home. I daresay she would even be less worn out.' With a tinge of resentment in his voice, he added, 'I'm on my own all winter with no one around.' He first worked as an engine operator at the local cooperative farm. In 1985 he was the first to obtain a business licence in the village. He supplemented the family income by repairing vehicle bodywork at night while he worked in his main job during the day. They also raised animals: four cows and a hundred pigs every year. Thus, he had had enough experience in managing a private business to be able to cope with any adverse impacts of the regime change. He bought a clapped-out combine in 1991, which he upgraded in his workshop at home. Then he gave up his job at the cooperative, which was on the verge of bankruptcy, and started to do contract work for the largest farmer in the village (since then Mr D has become the second largest). With the most financially convenient form of taxation he could find, he cultivates, as a so-called 'primary producer', an 87-hectare piece of land, which he had bought up piece by piece. The land is owned by his two daughters and himself. He set up another business, with his wife as its owner, with occasional contract work done with his vehicle fleet that had become significantly more sophisticated in the meantime, and animal husbandry as its main interests. When asked why the wife was the owner, both of them replied that you only had to pay a minimum amount of social security tax if you ran a private business that complemented your main job. Although Mr D often hires seasonal labourers to cultivate land, tending to the cows is Mrs D's core responsibility. Her working hours as a doctor's assistant exceed eight hours. Add to that the housework, and at least three hours taking care of the animals every morning, milking in the evening and occasionally hoeing she has a long day. Despite the heavy workload, she does not allow her career be interfered with. On the contrary, she has even applied for a place at a healthcare college.

> I'd made up my mind, and applied for a place [at the college]. There were no objections, but my husband couldn't help asking, 'Why do you want to do it?' He may have thought that I wanted to prove that I didn't have to depend on anybody for a living. You know what I mean. He thought maybe I wanted to break free, as it were, to prove that I was able to earn my own living. And I told him that he shouldn't get me wrong. I wasn't doing it for anybody's sake. I just wanted to see if I was able to do it.

When Setting up a Business was a Joint Decision

In the east, Mrs E had worked as a cashier at a cooperative farm. Her husband had started out as an agricultural worker, and had had several jobs. Made redundant in 1992, he was unemployed for six months. He was dismissed from his next job at a border post for a misdemeanour, thus depriving himself of eligibility for unemployment benefit. Like nearly everybody in the village, they also had a greenhouse, however, they regarded income from horticulture as supplementary. In mid-1990 they were able to purchase a small building, at a reasonable price, from the cooperative. This place was first operated just as an animal feed shop, then it was expanded to include video rental and confectionary sales. The business was run, on paper, by Mrs E. Her husband said that he would have thought about starting a business anyway, even if circumstances had not encouraged it. His wife, however, unambiguously attributed becoming self-employed to external constraints. If that had not been the case, she would not have thought about it at all.

No, [but for this] our life would have been more relaxed. I'd have done the usual chores before 8 a.m. anyway. You know the housework, the jobs around the house, the greenhouse; even if we had both had a job, we would have been able to manage things. Just like now that we have this business . . . I've always been wary of a business like this. It was my husband who was hell bent on it and talked me into it.

Still, the business licence was obtained for her so that she would not be unemployed and would have an unbroken employment record. Her husband admitted that he was afraid that 'they might find fault with something' if he had applied for a licence for himself because of that affair at the border station, though he now had a clean record. They share work, of course. Mrs E works in the shop serving customers. Her husband is responsible for acquiring the goods to stock the shop. Tending to the 10000 green pepper plants and the 3000 cucumber plants in the large greenhouse keeps both of them busy (Plate 6.2). Mr E admitted that his wife worked more, saying, 'She is more of an early riser than me.'

Despite Mrs E's initial worries, they both said that they were better off running the business than they were in their old jobs. It was more financially rewarding, and as Mr E put it, 'Nobody bosses me about. I'm my own master.' Mrs E shares her husband's opinion, saying that this lifestyle gives her 'more leeway'. She would not go back to being employed, she added.

When Setting up a Business was the Female Owner's Decision

Half of the female interviewees fell into this group as they independently decided to set up a business. Both in their early 50s, and living in the western border zone, Mrs F and Mrs G responded to family problems by starting their respective businesses. One problem was the downsizing of their husbands' workplaces, and the other was related to illnesses afflicting their husbands prior to or simultaneously with the downsizing. With their own homes doubling as business premises, Mrs F opened a dress and accessories shop in 1990, and Mrs G an ice-cream parlour in 1990.

As Mrs F recollected, 'My husband was taken ill. And I had to take things in hand, not bullying anybody, of course . . . I let him lead a more relaxed life, and looked after him for my children's sake.'

They started the business because her husband was having health problems. Their daughters were then aged five and seven and she was worried about having to support the family on only her salary. Her sister has a family business. She is a fancy leather goods maker and married into a family of self-employed people. In addition, both her father and her brother were involved in selling, her father selling wheat for the local cooperative farm and her brother being manager of a big department store. 'So I'd say we all had an inclination to do retail business.' At first the business was only part-time but after a while she was unable to cope with her two jobs and made the decision to give up her job in the food industry. 'I was the only one who could deal with purchasing the goods for the shop . . . if I gave up the enterprise we would surely lose what we had already invested when we started.' Thus, in 1993 she gave up her factory job and applied for a business licence.

Her husband had been a steelworker but like his wife he had completed his high school education. He now runs the menswear boutique. 'He is actually quite

Plate 6.2 Small shop in house with greenhouse behind in eastern border region

independent. I can trust his sense of style. He can do his own buying in the menswear part of the enterprise, which makes it easier for me.'

By contrast, running her private business as a supplementary economic activity, Mrs G still works as a financial clerk for a local company. Her husband has been registered as an unpaid family worker. In 1991, having heard rumours of layoffs, she had the idea of buying a second-hand ice-cream machine. They then had the machine installed in their home, and called it an ice-cream parlour. At first, the husband did not work there. Later, however, his place of work was privatized, and he lost his job. To make things worse, because of some infection that he had contracted, he was not allowed to perform heavy physical work. In his wife's opinion, 'he has been unable to develop a liking for this business'. But whether he likes it or not, his help is badly needed. Mrs G believes that they did not decide to start this business exclusively through economic necessity, but rather as a precaution and a safeguard.

Deciding to set up their businesses was a response to an unfavourable situation at the workplace rather than to the regime change in the case of Mrs H and Mrs I. These were the two interviewees who asked us not to interview their respective husbands.

Mrs H had been in charge of an old people's daycare centre for ten years by the early 1980s, a job that she liked very much, and she also thought that everybody was happy with her work performance. She became eligible for social housing. She was even given a key to her flat, which was, however, taken back from her the following day. 'The chief local councillor wanted to give the job to someone else, so I had to return the key. I handed in my notice.' Just a month before this Mrs H had married a divorcé, whose family owned the first pub to open in the village. The business that her husband ran in the neighbouring town was in his mother's name as he could not get a certificate of a clean record. After their marriage, Mrs H felt it was natural that she should take over her husband's business, at least on paper, and should work in the family pub. Before long, they had a restaurant cum guest house built. Although she already had a healthcare qualification, she also went to a catering school. Finally, the ownership of the whole business was transferred to her. Her job kept her so busy that it was her husband who stayed at home with their daughter. The restaurant on the upper floor was opened in 1987, and 'catering was my responsibility. Mónika was born in the same year. My husband stayed at home with her, as I was the only cook.' She felt running this business was something in which she 'got involved almost accidentally', but now she did not have any regrets.

By contrast, the career of 53-year-old Mrs J as an entrepreneur has been true self-fulfilment. She was the only interviewee with a university degree. She already had a degree in geology when she started to study for another one in road engineering, the latter being a requirement for her later job in north eastern Hungary. She got a divorce, and moved to Győr-Moson-Sopron County in the early 1990s. She started to work at the architect's office of the postal services. When she was, however, asked to make a network plan, an assignment for which she was not qualified, 'enough was enough'. 'I thought that I'd studied so many different things, and that if I couldn't rely on them, I didn't want to study anything else.' She gave up her job at the post office, becoming the first registered unemployed person in the city. A year passed. She bought a video recorder with money she had inherited from her mother, and enrolled for a training course in video recording. In 1992 she started a business. Today she is a successful producer of nature films for educational purposes. The initial years were very hard.

'Back in the socialist era we weren't taught to stand on our own two feet. No one ever taught us how to. Ever.' She said she did not know from whom she had inherited a knack for running a business. 'I'm determined to live a meaningful life and to leave some legacy. And if this is what I have to do in order to be able to do so, then I do this. Although she seems to be able to manage her business quite well on her own, her marriage has suffered.

> I think it's safe to say that running a business changes your life dramatically even if you don't have any children. I don't have any, but if I had, this would be even more true. It changes personal relationships. That is for sure. For if you're not business partners, if your partner's not involved in the daily running of the business, doesn't share problems with you or provide support for you, but rather, you compete with each other, then it destroys everything. Especially, if you do reasonably well, but your partner does not. My husband, for instance, is a public servant. He's convinced that he has to work a lot. In fact, much more than I do, he will add. I'm sure he is right from his own perspective. 'It's easy for you because you can do as you like.' What I really like doing, and how it is related to what I actually do is a different story altogether, though. I never mind doing what I have to, though it's true that I try to find jobs for which I'm cut out. I must also get some spiritual support from work.

In respect of gender relations, Mrs K's story reveals characteristics other than the ones described above, providing an example of possible new forms of such relations. At 37 years of age, and living in the west, she was in her second marriage. Her husband was also divorced. She had stayed at home with their three children for nine years, receiving childcare allowance. Then she joined a German firm that assembled radio spare parts. She commuted to the neighbouring village daily. It soon turned out that, compared to the work schedules at state-owned companies,

> here you never knew when you'd be allowed to go home. If an order was placed, it had to be delivered . . . Didn't matter if you had children to look after. If the deadline was tight, you had to stay on, no matter that there was no shift work. There were times when you only managed to catch the midnight bus, and then you were practically never at home.

It is no coincidence that, after this work experience, Mrs K represents those women in the sample who stress the importance of the family, and appreciates that self-employment allows her to stay at home. 'The reason why I chose self-employment was that when I was no longer eligible for childcare allowance, when my daughter was nine, I thought I could spend more time with my family than if I commuted to work.'

Naturally, they also needed the income from the business. Mr K had had several jobs and been unemployed on two occasions before he finally found a job in an aluminium factory in a neighbouring town in 1994. He thought the pay was good, but his wife did not. Although they went over to Austria to do seasonal work, what he earned was rapidly spent on house construction. The idea and the possibility of starting a business arose when the local cooperative farm decided to sell its cattle herd and lease out its milk depot. Mrs K purchased the building in 1996, using her compensation vouchers and those of her parents' issued by the cooperative, and taking out a special loan for entrepreneurs. After the initial difficulties had been overcome, she added animal feed and cooking gas distribution to the business. Her husband supported the idea of starting a business, urging his wife to 'try her hand at it'. Retrospectively, they both think it was a good move. In addition to the financial rewards, said Mrs K, 'now I have more free time, and I can schedule my own time'.

Not unlike many of our male interviewees, her husband singled out one advantage to his wife's running a business. 'Overall, I was happier about it than not. She didn't have to go out to work. Her business kept her here. The children still go to school. It does give you peace of mind to know that everything's alright at home.'

He said that in the beginning he had helped out, but that there was no need for that now. Mrs K said that when her family helped out with the sale of animal feed, given that they were not familiar with the goods or the prices, 'there is always chaos', which she did not like. According to Mr K, his wife is the sole decision-maker, 'she's at the helm'. He says that she always makes the right decision. Mrs K, however, said that this was an expedient standpoint for him to adopt. In her opinion, 'the only thing that my husband cares about is to have everything. He doesn't care how.' She felt she was entitled to her decisions. She went on to explain why.

> Providing financial support for the family is mostly my responsibility, since he chips in with very little. Because of repeated rows over his grown-up children [from his previous marriage], we no longer have a joint family budget. Everything is my responsibility. I pay the overheads. His bank account is only debited with the amount of the monthly instalment of the mortgage. He pays nothing else . . . So once I decide to do something, then I'll do it. I take care of the animals at home, too, because, except for feeding them or cleaning their sties sometimes, he can't be bothered. I keep pigs. There are sows and I always raise the piglets. And when I sell them I won't have him telling me what I should spend the money on. We've just had another row over it. It's alright that he gives some to his daughter [from his previous marriage] from the money we got for this porker. But in that case he really ought to chip in when it comes to buying the feed.

As regards the underlying reasons for female self-employment, the above excerpts from the life stories of our interviewees attest to the importance of a few such factors that other studies have also pointed out. Although our sample included businesses set up during the socialist era, indicating the continuing importance of a family legacy of entrepreneurial experience, the social and economic transition was unequivocally among the major causes encouraging self-employment. With the emergence of the market economy, many people felt that setting up a business of their own could improve their lives and that of their families (Gere, 1996). It cannot be gainsaid, however, that high unemployment was also a contributing factor, leaving no choice to many but to start a business (Gere, 1996; Nagy, 1996). The economic problems of the transition impacted the family members of the majority of our interviewees, whether they were commuters or worked locally. At the same time, however, a large number were able to use, as seed capital, compensation vouchers or real property or machinery bought cheaply during the transformation of cooperative farms or the liquidity proceedings of companies. Coming from a family with an entrepreneurial background also helped women to think positively about starting a business. When the business was started at the husband's initiative because of his family experience then the wives were more worried, although they usually were able to accept the risk of self-employment eventually.

Other studies have not investigated the role of gender relations in the emergence of businesses. It is not always because of the woman's wishes that she becomes an entrepreneur. Sometimes they turn their husbands' dreams into reality. One thing is sure, however, gender relations in the family do play some role in the emergence of businesses. Nagy (1996, p. 64), having conducted interviews with some successful female entrepreneurs from Budapest and a few from the provinces, concluded that 'a supportive family background was a sine qua non of a successful business'. She defined supportive in the sense that the family lends an understanding ear to 'women's efforts to assert their independence'. Such a background was, however, missing in the case of many of our interviewees, who were, for the most part, also successful. There are a number of examples of a husband's insistence on his wife's giving up the job that she liked 'for the sake of the business', and of a woman for whom running a business does mean self-fulfilment, despite friction within her marriage. Nagy (1997) noted that a higher proportion of her successful women entrepreneurs were divorced than she found among men entrepreneurs. However, we had cases among our interviewees where both husbands and wives had supported the setting up of a business by the spouse and even when initially not supportive had become so over time.

No matter who is responsible for the final decision, the relationship between female entrepreneurs and their husbands seems to affect, directly or indirectly, their decision about starting a business. Gender relations on the household scale are unequivocally interwoven with those on a national scale, in terms of lower pay levels for women compared to men, and the patriarchal gender roles that are accepted on a macro-social level, especially in rural areas.

Living in a rural area, the businesses that our interviewees ran were generally carried out, in part or wholly, in the home or nearby (see Table 5.3). As a result of the combined effect of the rural scale and specific gender relations, an overwhelming majority of the husbands perceive their wives' businesses as allowing them to stay at home, tending to traditional female duties. Many of the women also appreciated being able to give up commuting and to be at home with their children. They also liked the flexibility of organizing their own time which was impossible when they were employees.

The actual content of the gendered nature of women's work as entrepreneurs in the western border region seems to be different from that in the eastern one. Given the small size of the sample, no far-reaching conclusions can be drawn about this issue although it is supported by our wider questionnaire survey.

Opinions of Entrepreneurs on Space and Place

Village versus Town

As regards the evaluation of the characteristics of a residential environment as well as spatial differences and inequalities, three geographical scales and places cropped up as terms of reference in the narratives of our interviewees. One is the village. Many voiced objections to the village as such rather than the respective villages in which they lived, citing disadvantages, undoubtedly more numerous in the south-eastern border region than in the north-western one, compared to the town.

Three of our interviewees from the former area replied to our question about their business development plans by asking a question. 'What's there to do in a village?' Mr A found that the range of possible business activities was limited. Animal husbandry and whitewash distribution (his business activities) and the animal feed business (in his wife's name) all rely on local farming as their market and source of grain. They deliver feed to shops in seven neighbouring villages and have their own shop in their village so reaching a wider market. Mrs A has four male employees who do the deliveries and work in the local shop. Mrs A pointed to low demand when we asked why she did not open a restaurant where she would have been able to do what she had been trained for. 'There is no demand for that. None at all. There is no such thing here as tourists as in the town.' Mrs C mentioned both limited possibilities and low demand in her reply. After the failure of the various farming activities in which they were involved and the second-hand clothes shop that they ran, she felt that the only possibility left was to open a 'dime store' that sells cheap goods.

Unemployment was also mentioned as a serious issue in this region by two of our interviewees. According to Mr C, 80 per cent of the population in his village, where about 1500 people lived, did not have a job. This, in turn, means that, though of very good quality, the feed that he sells in several villages is very expensive for 'the villagers here'.

Mr D also said that only 15 people had jobs in his village. Like others, he also pointed out that this was the outcome of deteriorating economic conditions, adding that the local cooperative farm used to employ 50 people. Remembering his career as a self-employed carpenter, Mr A said that there used to be as many as 14 carpenters in his village when he was young, 'all had been kept busy day and night'. 'They built as many as 40 houses every year. Not any more.' For him, harsher economic conditions are reflected in these changes. Although Mrs K, owner of a milk collection centre, who lives in the north western border region, also mentioned that her business had been affected by the fact that there were now only six persons who raised cattle, compared to the 16 when she had started her business, she attributed this decline to the ageing of the population. 'The young decide to stay only if they inherit a house here.' Dwindling turnover also poses a problem to Mrs H, the local restaurateur in the west. She said that the main underlying reason for this was that '[local] families cannot afford to dine out on Fridays or Saturdays like they used to when we opened this place. There were times when people queued up for fish soup on Fridays. Back then, we cooked 120–130 portions, now 20 at best.'

Although also complaining about the above two phenomena, our interviewees in the western border regions also mentioned some progress, although certain aspects of this progress were not necessarily advantageous to them. They perceived the fact that 'businesses have mushroomed' as increasing competition. As quoted above, Mrs H shared this opinion, saying that a large number of restaurants had been opened. There is no contradiction between her two statements. When her restaurant started before the transition there was no competition and it was a seller's market.

It is mainly due to the growth of economic polarization that envy is rife. All our interviewees in eastern Hungary and one in the western border region said that envy, often targeting them as successful entrepreneurs, was a growing sentiment in their villages. As Mrs E put it, 'if you run a business, you're envied'. Mrs C said that the reason why their second-hand clothes shop did not do well was that not even their neighbours did their shopping there. 'To come to our shop? Never! They'd sooner go over to Gyula [a town nearby] and fritter away their whole day than come to this shop.' However, in the western border zone most shops depended on local customers. Mrs F said that despite being near the border she only had about one foreign customer a week. She felt that she would need to open a shop in a tourist area to get more foreign customers but the rent would be high so she had decided not to expand outside her village. The reason why Mr D does not employ illegal workers from Romania is that he is afraid that somebody in the village would turn him in to the authorities as they did in the socialist era when he first obtained his business licence and repaired cars at night. He was turned in twice, but everything was found to be above board. The interviewees all agreed that these manifestations of jealousy did not come from fellow business persons in their respective villages. On the whole, they got along quite well as a group. This suggests that social polarization goes in conjunction with a certain social tension.

Despite the numerous drawbacks to the village that they mentioned, our interviewees, especially those living in the east, kept emphasizing that the town, representing the other end of the spectrum, was not necessarily more attractive. Mrs A, who has lived in a village all her life, said that she would not be able to get used to living in a town. Mr C, who moved from a town to his natal village, sees some justification for a not too successful change in his life in the fact that 'life is quieter in the country and the air is cleaner, too'. Unlike Mr C, who cited the advantages offered by the environment, Mr and Mrs E referred to their own hard work. As Mrs E put it, 'urban folks don't have the slightest idea of how hard we can work'. Her husband did not mince his words when it came to urban lifestyle. 'We're not accustomed to what folks that live in blocks of flats are. They go home, watch TV or walk round the town. We're peasants. We're accustomed to hard work.'

Regional Inequalities – East–West Relations

Placing the advantages or disadvantages of the village in which they live in a wider context, a few refer to them as regional characteristics. One is Mr F, in the west, who perceived an increase in the number of businesses as a characteristic that both his village and the neighbouring areas shared. Unlike him, Mr A, who lives in the South Eastern border region, said that 'we live in the wrong place'. Also including the neighbouring villages in this 'place', he summed up his opinion of them rather disparagingly, 'This place sucks.'

Our interviewees, irrespective of where they lived, also referred to the East–West differences. Mrs J, the entrepreneur with a university education, though she could not fail to notice poverty in the Great Plain and predicted an even bleaker future for the eastern region, also set great store by differences in attitudes. She perceived some of them as regime-specific. 'Communism did not brainwash the folks 40 to 50 kilometres from here across the border [i.e. in Austria]. And their attitude has been adopted here. This is why there is West HERE, and East THERE.' By 'there', i.e. the 'East' she meant places like the two hubs of heavy industry (one in Eastern Transdanubia, the other in Northern Hungary), 'where communism was instilled into folks' minds, and where life depended on the mine, which they took away. And now the folks there have been left high and dry. And they can't see any way out. Now, that's a fine kettle of fish there.'

Living quite close to that 'fine kettle of fish', Mr C also thought they were in a dire situation, saying that the future was rather hopeless. His opinion, which was similar to one that also appeared in print, was summed up as follows:

> development will leapfrog the South Eastern region when it finally gets here, with investors setting up their next base in the Balkans. Unfortunately, EU accession comes from THAT direction. If Hungary accedes to the EU, we'll be the last to notice it. Nobody comes here. Everything and everybody stops at the River Tisza. The soil is of poor quality, there's no infrastructure, no roads, nothing at all. The cities in West Transdanubia, Transdanubia and maybe the region between the Danube and the Tisza are still accessible. There're airports there. Here we've got nothing, only low quality soil. Western Europeans don't want this, or marshland or aspen groves. There's nothing in it for them. The same with village tourism here. This is backwater . . . a ditch that is jumped over.

There is more to this comparison between the West and the East than geographical unevenness. Symbolized by the European Union for Mr C, all that is good can 'dissolve national boundaries'. He perceives the River Tisza as a border that cannot be crossed, with the 'virtually redlined' territories of global economy to the east of the river 'experiencing the most intensely debilitating effects of uneven development' (Smith, 2000, p. 868). These territories are 'a place of total exclusion', where everything is 'bad or poor', including soil, despite the fact that the best soil in Hungary can be found a few miles to the west of his village.

Only Mr A, also living in an eastern border village, offered a more subtle picture of the West. However, one of his sentences was a generalization even if it differed from common stereotypes. 'All this came from the West, drugs, computers, videos.' He did not hide his dislike for them. He had visited The Netherlands and Germany in the early 1990s, and came back with the belief that 'there are a lot of poor devils there, too' imprinted on his mental map rather than seeing the possibilities of life in Western Europe.

Perceived Differences between the Two Border Regions

In the narratives the most marked spatial differences and, in part, inequalities are associated with the two study areas as border regions. Though not including his own business, Mr B, who lives along the north-western border, stressed that those businesses that exported had become prosperous, owing to the proximity of the international frontier. By contrast, Mr A and Mr C in south-eastern Hungary insisted that no business could be based on the Romanian market.

> There's a border-crossing point near here. If you need something, you cross the border and buy it. They're not that disadvantaged over there. They can buy anything if they have money. As to Romanian goods, except for a few things, like spirits, wood and kitchenware, they're not in demand here. The market is so saturated that entry's almost impossible.

Perhaps the most striking difference lies in cross-border employment, which also leads to uneven living standards. Sík's calculations (1999) reveal that the proportion of Hungarians in employment abroad is the highest along the Austrian border, in terms of both frequency and intensity. Several of our interviewees in the Győr-Moson-Sopron border villages mentioned their frequent employment in Austria. Mr K admitted that if they had been able to speak German, they would have opted for permanent employment there rather than setting up a business. As they did not speak German, the only possibility left for them was working in the annual grape harvest. Mass job-seeking in Western Europe may even hit our interviewees adversely. One of Mrs B's employees left for the sake of a more lucrative job in Austria. Mrs J also perceived this situation, though, for the time being, it did not affect her personally, as a source of several problems.

> . . . labour here along the border is drained. Unless I can make employment worth their while, no one will work for me, no matter how rewarding the job I can offer is. They'd sooner cross the border to slave away at some job, cleaning houses, picking grapes, anything. But then, they also need the money. You understand them. There's a huge income gap, at least here. If I request cleaning services, they charge me more because of the vicinity of the border, for the sole reason that you're in the western part of the country.

Precisely the opposite is the case in the south-eastern border region. Employed illegally in some villages in Békés County, Romanians from across the border are seen as a source of cheap labour. Information available to us and confirmed by our interviewees, was that Romanians were mostly employed in two or three villages where greenhouse-based horticulture is the mainstay of the local economy. The largest of these villages is Méhkerék, the country's only village where ethnic Romanians constitute the majority of the population, making employment easier for both Hungarian and Romanian speakers. An ethnic Romanian himself, Mr E said he did not employ any Romanians, but he was in favour of their employment. As he put it, it was not worth employing day-labourers 'legally', so 'you have to hire black labour sometimes'. At the same time, however, he went into a detailed explanation as to the dark side of black work such as heavy fines and police abuse. As regards the latter, complaints were made to the Hungarian Association of Ethnic Romanians.

> Complaints had to be lodged because they did such things that . . . They didn't come to my place. They'd better not . . . I'm completely outraged, and so is the whole village. What they did was that there was this peasant, who lived locally, walking along the street. They didn't so much as ask who he was but just grabbed him, handcuffed him, shoved him into the police car and took him to Sarkad [a town nearby] for questioning. They finally found out that he lived in Méhkerék, and was on his way home. They beat them up, the Romanians. I once worked at the border station, so I know what that means when Romanians cross the border. They've got their documents in order, so they cross the border, and they can stay here for 30 days without anybody bothering them. So there they are walking along the street when a policeman grabs them, shoves them into the police car and gets an expulsion order for them so that they're not allowed to enter the country for two or three years. On what grounds, I'm asking?

None of our interviewees in Győr-Moson-Sopron County referred to shopping tourism, a highly profitable activity in the late 1980s, as one of the advantages of living along the Austrian border, which tallied with the findings of other studies (Wastl-Walter and Váradi, 2004). Hardi's studies (1999) conducted on both sides of the Austrian–Hungarian border confirm that employment is the main reason that Hungarians cross the border. Austrians, however, mainly associate Hungary with the possibility of shopping or getting cheap services such as dentistry or car repair. Depending on the size of their businesses, our interviewees benefit from the latter phenomenon to a varying degree. Some earned extra income from jobs related to house purchases or construction by Austrians in Hungary. In Békés County just the opposite is the case in this respect as well. Car owners regularly cross the border to Romania to fill up their tanks as petrol is cheaper in Romania.

Finally, the manner in which the way of thinking of those living in the border regions reflects open or closed borders is another key issue. The reason why this is important is that, given the current political and economic environment, in addition to porous borders, the mindscapes of people may also interact with the intensity of cross-border relations. Studies conducted in a village on both sides of the western border revealed that, as a result of earlier geopolitical changes and separation, 'the memories of common history get lost' on both sides. In addition, recent political and economic changes and power relations also added to differences in cultural identities (Wastl-Walter and Váradi, 2004, pp. 188, 190), and, according to Hardi (1999), even new identities with components of discrimination and prejudices have emerged. They agree, however, that these changes may re-arrange cross-border relations. Our study identified not simply division but also sharp conflicts along the Hungarian–Romanian border. Manifestations of these conflicts between Hungarians and Romanians, also heard in several of the narratives, included utterances such as 'I don't want Romania' and 'Ceausescu settled migrant gypsies from Walachia in pre-Trianon Hungarian territories'.

Mr and Mrs E, who keep in touch with their relatives in Romania, are both ethnic Romanians from Méhkerék, where the majority of the population are also ethnic Romanians. Mrs E told us how she saw the consequences of such an attitude.

> There are mixed marriages among those who went to college or university. But, we're steered to marriages where both parties are Romanians. Because there've always been conflicts between Romanians and Hungarians, I grant that, that's true. Wherever we go, we are reminded that 'You're a Romanian' often in derogatory terms. You know what? You walk through the village and you'll see how hard-working the people here are and what they can do. And that you should've seen what the village like in the old times. No one that sets foot in the village can help saying that you can't find a similar village in the area. Only perhaps in Transdanubia, where conditions are different . . . But here, where . . . People are indeed clean, neat and tidy.

Mr C described popular attitudes in his village with an all-Hungarian population, where no Romanians were employed illegally, or only 'one or two occasionally', because there were no employment opportunities even for local people.

> Local people still feel resentment because of Trianon. Hostility still prevails between Romanians and Hungarians. What matters is not that somebody is Hungarian [from Romania], but rather, that they're from Romania, that is ours. For the people here, myself included perhaps, it's still important that it belongs to us. I'm not prejudiced, I mean no harm to anybody, but if someone asks me where 'ours' ends, I'll say to them, 'Under law here, in spirit over there.'

The sharp territorial distinction between 'ours' and 'theirs' and the difference between the de facto state frontier and the mental border that lies farther away to the south-east suggest the existence of a divisive mental geography that definitely does not help to improve actual cross-border relations. Overall, the way of thinking of our interviewees clearly reflects uneven economic development in the villages in the north-western and south-eastern border regions that represent the two extremes of contemporary Hungarian rural development. This holds true even if, from the perspective of small businesses, both advantages and disadvantages present themselves in a more subtle way in their day-to-day operations, than the wider lower-status strata of village population experience them (Timar, 2001).

'Inside' and 'Outside': Borderlines in the Gender Division of Labour

As was discussed above, uneven spatial development influences the gendered production and development possibilities of enterprises as well as the living conditions of the families of entrepreneurs no matter whether the owners are women or men. The question is whether or not spatial inequalities affect the gender division of housework, another important indicator of gender relations at the household level.

The narratives of our male and female interviewees clearly substantiate the fact that the traditional division of work prevails in their families. Division of work is still often associated with distinct territories of the micro-space of the home, with a sharp border-line lying between feminine housework 'inside' and masculine work 'outside'. This perception, the true interpretation of which is mainly possible in a rural context, was present in both study areas, though in the South Eastern border region every women participated actively, sometimes even more actively than men did, in animal husbandry and horticulture, work taking place outside the house. By contrast, only a few men did housework, and very little at that. While men rarely bothered to find an excuse for their non-participation, their wives sometimes did so for them. 'I can manage here inside, there's enough for him to do outside', said Mrs C. Although women did not deem such division of labour as a normal course of things, they sounded as much resigned to it as they sounded plaintive. As Mrs D put it, 'I've already given up'. Before 'giving up' she had had several conflicts with her husband.

Women's new role as entrepreneurs has not brought about material changes in this respect. None at all for Mrs E: 'everything is as usual. I have to remind him to empty the dustbin. He will shove things in no matter how full it is. When I remind him to empty it, he does so without a word of complaint. But he always has to be reminded. He takes it for granted that I take care of it.'

It was often the case that, after a while, even occasional help from men was discontinued. Such help had mostly to do with looking after children. This simply meant that men, their work schedules permitting, helped out when women, still with a steady job, could not manage all their tasks because of shift work or commuting. When, for instance, Mrs B swapped her job involving shift work for her husband's enterprise, which meant very hard work for her, 'in addition to this, there was the job of running the household. Everything had become my responsibility, as my husband went full steam ahead with his business.'

As has already been pointed out, the reason why some of the male interviewees insisted on their wives starting a business was that they wanted them to stay at home. And for one of the female interviewees this was also the main motivation. Her fellow interviewees also conceded that at least they could schedule their own time, which made their lives somewhat easier, and that they can run their respective businesses from their homes. Mrs A said that because of new regulations they were no longer allowed to have the pigsties and a feed shop on the same premises. Therefore, they had bought an old peasant house, across the road from their house, where they had built a modern piggery. Houses were very cheap in the village but Mrs A felt that the move would increase her work.

This one example suggests that combining reproductive and productive work may lead to the re-emergence of spatial conflicts if the business has to be run at places other than the home. An example of such a scenario is the clothes shop run by Mrs F in the west. She has to buy goods for the shops, which keeps her busy until late at night two or three times a week. Being away from her children was emotionally trying when they were young. However, she thought it was her own role as an entrepreneur, which she liked, rather than the unequal division of labour at home that was at the root of the problem. 'This is not a nine-to-five job. It's true that it's too much for a woman. Especially, if she has children.' Overall, it seems that in these villages having a home-based business only reduces women's frustration, but not the amount of housework that they have to work.

Our interviewees were conditioned to traditional female roles under socialism when they were both expected to have a paid job and to cope with reproductive tasks but were able to get time off to look after children. Their parents were born before World War II, however, and their married life coincided with the socialist era. Thus, most mothers had to work on cooperative farms or elsewhere. Meanwhile, they also usually kept animals and cultivated family allotments. Those who decided to stay at home mostly did so after their children were born, as often there was no public childcare in a number of villages. Nevertheless, when talking about their mothers, our interviewees (men and women alike) often said that they 'didn't work', in fact more often in reference to their mothers' lives than to their own. Mr K's life story shed light on familial patriarchy. 'In our family my father was head of the family, and my mother had to do as she was told.' Or as Mrs C put it, 'My stepfather was strait-laced, and would often say that men were masters of the family.'

Quite a few labelled the division of labour between their parents as '*traditional*'. The pattern had been passed down from parent to child. However, when they compared their own lives to their parents' lives, differences in male and female perspectives were immediately obvious. Several women recalled the impact of unequal gender relations on their upbringing. Mrs D, for instance, hinted at strictly separated roles. 'My brother had to focus on the jobs my father did . . . he always had to learn what my father was doing. It never occurred to him to wash up a dish. He didn't have to learn that because that was not "a boy's business".' Mrs K mentioned even greater differences. 'I'm not saying that I only had to do what girls are expected to because I also had to help round the house. My brother didn't have to help inside. I had to learn everything both inside and outside.'

Undoubtedly, some of the men in rural societies had to start working at an early age. However, they only complained about or were proud of their hard work. Mr C, for instance, herded '25 geese along the village boundaries' when he was four years old. None of them referred to having been treated as inferior to their sisters. It was often the case that they adopted the pattern they had inherited from their parents to their own life. Some of them found it especially important to emphasize this when they sensed that they had to explain their attitudes to housework. As Mr F put it, 'Well . . . doing the washing up? I had also helped with it before the business was started. But a man to do the washing up! This was unheard of at my parents' place.'

Mrs E's story confirms that even if the pattern passed down from parents differs from the traditional perception of gender roles, there is a chance that men's adult environment transforms this pattern in a way that is more adjusted to the macro-social norms that are more beneficial to men.

> My mother-in-law gave my husband a good family education. She taught him how to behave towards women. You can see the difference between my father-in-law and my husband. That a woman is weaker or something. But, how shall I put it, as life passes, you swim with the tide, and what you soon realise is that your neighbour never does such work, and that it's a woman's job. But that's not true. I know for a fact that my mother-in-law taught him to be nice and polite to women, but, you see, the nature of men cannot be changed.

Even those in favour of a different division of labour relegate gender roles deeply rooted in society to the category of 'human nature' only too easily, thereby transforming social gender into biological sex.

It seems that cultural traditions thwart the adoption of even relatively good practice. Ethnic Romanian Mrs E from Méhkerék, who referred to conflicts between Romanians and Hungarians, was quick to identify and ready to adopt what was good practice of Hungarians, to no avail. She attributed her failure to do so to her ethnicity.

> Doing the housework? . . . Men rarely help with it here . . . In marriages where both parties are of Hungarian origin husband and wife share housework and the responsibility of raising the children, including changing the nappy and everything. There are mixed marriages here, too. The husband gets up in the night to feed the child and change its nappy. Lads from Méhkerék hardly do such a thing. I think this is because there's hard work here all through the day. What your job is matters a lot. If you go out to an eight-hour job it is easier.

Essentially, the village seems to intermediate traditional gender roles.

> You won't see a man with a hoe in his hands. They're used to women's doing hoeing . . . They think it's a woman's job . . . Men will say that work round the house is a woman's job.
> [Mrs E]

> No, offhand, I can only remember ten men at best, who lifted this burden from their wives' shoulders. Unfortunately, this is still the case in the villages.
> [Mrs D]

This finding of our studies is very likely to apply also to places other than the villages in our sample area. A survey (involving 1023 persons in Hungary) conducted as part of the International Social Survey Programme in 2002 revealed that women spent a national weekly average of 27.7 hours doing housework, compared to men's 11.0 hours. A full breakdown of these data clearly reflects settlement hierarchy: women in the capital city and county seats spent a national weekly average of 23.3 and 23.2 hours respectively doing housework, while those in other towns and villages did 27 and 34.2 hours respectively (Blaskó, 2002). The differences between villages and cities may reflect the greater availability of modern household equipment in urban areas as well as the absence of animals to feed and the more conservative attitudes in the villages.

The future may bring changes as it has elsewhere among the younger generation. Mrs D, who made efforts to provide modern family education for their children, had long abandoned her hopes for her son. Her daughter, however, wanted to break with current traditions, which she criticized severely.

> Once when I complained, or rather moaned, my daughter told me that 'my marriage won't start like this. From the outset, we'll share work, and that's what he'll get used to, not this.' She says, 'It's your fault because you let Dad have everything his own way.' I do everything, and he has nothing to do inside the house.

During the 'full employment' of socialism, our female interviewees had to learn how to combine career with family. As other research confirms, the efforts made to reconcile the traditional with the modern were not necessarily an example to be followed by children, i.e. the next generation (Neményi and Kende, 1999; Timar, 2004). The husbands of our interviewees mostly experienced the disadvantages of this era for family life. Given this experience, and with wives who used to have a steady job and most of whom had a certificate of secondary education, they sounded uncertain as to what an 'ideal' world would be like. They also said somewhat tentatively that 'women shouldn't go out to work'. In the same breath, however, they also added that if that were the case, 'they'd be bored'. Most deemed female enterprises a happy medium, since this meant paid work that prevented female 'idleness', but also kept them at home.

Conclusion

Overall, entrepreneurship in villages has not changed the traditional division of labour. In fact, since it involves women working from their homes and with more flexible hours than when they had worked as employees under socialism, it may have allowed men to feel that there was less need for their help in the house. In border regions, whether developed or not, we experience the same gender roles and attitudes in most families. In some cases children helped out in the home and some husbands also did so but probably did not want to talk about it as it did not fit the generally accepted pattern of gender roles. What does exert an impact is the rural environment and possibly ethnicity.

Chapter 7

Conclusion

Economic restructuring is a regular occurrence in rural areas as non-renewable resources are worked out or the conditions for agricultural production change. It is particularly difficult to deal with in these areas as rural labour markets provide fewer opportunities because of sparse population and spatial constraints on travel (Hardill, 1998). New rural survival strategies have to be developed combining both reduced household consumption as well as new sources of income (Tykkyläinen et al., 1998). For young people, the survival strategy often takes the form of migration to urban areas leaving behind a village population dominated by the elderly, women and those least likely to take risks. Thus innovative development in rural areas can be difficult.

Rural restructuring in the transition countries has followed a similar path to that in the rest of the world except that it was precipitated more suddenly by political events in 1989/90 and the rural areas involved had characteristics not found outside the centrally planned economies (Paul, 1992). The impact is further exacerbated by the expansion of globalization, neo-liberal economic policies and the accession of many of the transition countries to the European Union in May 2004.

The effect of these rapid changes varies, with some peripheral areas benefiting from border changes, as in north-west Hungary, while others remain peripheral because of their unfavourable location or socio-economic structure (Paul, 1992). Smith and Pickles (1998, p. 5) also noted the diversity of forms taken by the post-socialist transition and their impact: '[T]hese changes have in turn brought with them a reworking of the geographies of the region in the forms of metropolitan growth, the economic collapse of peripheral regions, the polarisation of urban [and rural] spaces as inequalities deepen, and the reworking of the territorial structure and democratic spaces of the state and civil society.'

In our study we looked at two isolated rural border areas and the role of micro-businesses in coping with these gendered changes in the labour market. In the last year the western border has become more open and the eastern border more closed following Hungary's accession to the European Union. These political changes are reinforcing the regional differences we saw in our study. The western border zone is rapidly becoming a post-productivist rural area with tourism increasingly important. In the east rural areas remain largely agricultural. Thus opportunities for small businesses are becoming more plentiful in the west and less so in the east. Between 1996 and 2003 the population of Hungary as a whole declined, as did the population of Békés County in the east but in the west the population of Győr-Moson-Sopron County grew. In the border villages in both areas those villages with newly opened road border crossings are growing while most others are losing population. This does suggest that border crossings are becoming points of attraction. In the west out of 19 border villages, two of which had not been included in our study because we could not obtain lists of their entrepreneurs from the local government offices, 11 had increased their population between 1996 and 2003. Of these, nine were located close to the National Park indicating the growing importance of tourism in this border region.

The growth of small businesses also differed in the two areas. Between 1996 and 2003 the number of small businesses in Hungary grew by 2.8 per cent. In Győr-Moson-Sopron the growth was 15 per cent and in the western border villages, 14.6 per cent. Clearly this western gateway county was fertile ground for entrepreneurship but more unusual was that the small rural border villages had virtually the same growth rate as the county as a whole in terms of new businesses. These figures support the idea of the western borderlands being areas of dynamism and innovation with a flexible and mobile workforce. In the east Békés County had a decline of 0.9 per cent in the number of registered businesses but in the border villages entrepreneurial activity declined more precipitously by 7 per cent. Thus even before accession to the European Union the western border villages were growth areas for entrepreneurs while in the east the border villages were far worse than the county as a whole, although both showed negative growth in businesses over this period (Figure 7.1).

The overall pattern is gendered and has led to major changes in labour markets, as has been noted elsewhere (Beneria, 2003). These changes involve increased flexibilization of labour, a widening gender wage gap, marked polarization of incomes, instability in labour markets, a greater dependence on contingent work and less social protection. Becoming self-employed involves far more risk than was common under communism when there was little choice of employment. Some of the businesses that are officially licensed to women may be so to avoid men having to face the risk of failure and the associated shame. However, the explanations we were given for this choice of job were overwhelmingly related to taking advantage of a particular tax regime or maintaining eligibility for social security and pensions or helping the family. Yet while providing these economic explanations it was quite clear that for many women being an entrepreneur was exciting and enjoyable. Despite complaints about the negative effect of post-communist changes, the vast majority of our respondents did not want to return to being employees. This positive reaction was true in both areas, even in the eastern border zone where pessimism was widespread. Most of the entrepreneurs we have studied are involved in micro-enterprises with none or very few employees and many did not want to expand as it would mean adding employees which they did not want to do because of distrust of non-family members and also because they did not want to take on the extra paperwork required.

Self-employment allows people to have the freedom to take advantage of entitlements and capabilities (Sen, 1999). Under socialism the maternity and childcare allowances available made women responsible for childcare and so reinforced gendered divisions of labour. Discarding stereotypes about men's and women's jobs provides both new opportunities and increased friction between men and women. Yet sexism is increasingly being used to prevent women moving into better jobs. Our research showed that men and women entrepreneurs have different attitudes to their work. On the whole male entrepreneurs went into business to make money. For women there were two approaches depending on whether they chose to become self-employed or were forced into it because of economic necessity. For this latter group women saw entrepreneurship as a means of family survival and as supporting their husbands, while those choosing to become self-employed saw their business in terms of service

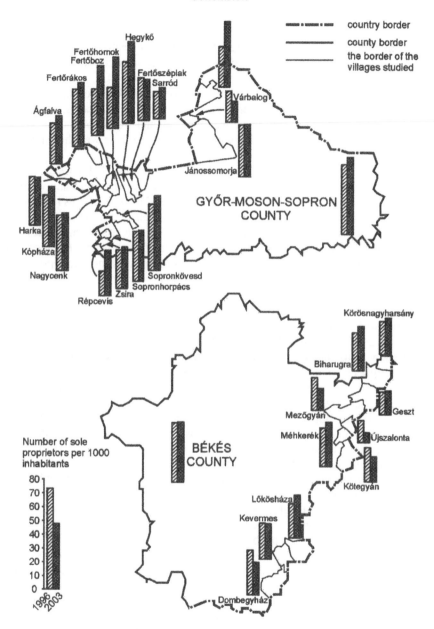

Figure 7.1   Number of active sole proprietors for villages studied, 1996 and 2003

Sources: Statistical Yearbook for Békes County 1996, Hungarian Central Statistical Office, Békés County Management, 1997. Statistical Yearbook for Békés County 2003, Hungarian Central Statistical Office, Békés County Management, 2004. Statistical Yearbook, Győr-Moson-Sopron County 1996, Hungarian Central Statistical Office, Győr-Moson-Sopron County Management, 1997. Statistical Yearbook, Győr-Moson-Sopron County 2003, Hungarian Central Statistical Office, Győr-Moson-Sopron County Management, 2004.

to the community and as a means of self-fulfilment. Thus entrepreneurship either reinforces the dominant patriarchal culture or creates a counter-cultural discourse on women's role in civil society.

The success of many women entrepreneurs leads to increased relationship problems in families. Women are adapting more quickly by learning languages, especially German, and by choosing to stay home with children and work from home. These changes in lifestyle often bring friction within families and there were high rates of divorce in entrepreneurial families.

Differences in attitudes among women entrepreneurs, as revealed in focus groups and interviews, also reflect variation in levels of living on the eastern and western borders. In the east the stress of poverty, uncertainty and the loss of jobs resulted in increased levels of depression and general poor health. Men take disability retirement to reduce the stigma of unemployment but women do it much less often. Kovács and Váradi (2000), in their study in a small rural town, found that class and education and the related social networks were the keys to success for many women entrepreneurs. We also found the same indicators but they needed to be supported by determination and a willingness to take risks.

The spread of entrepreneurship in villages is increasing economic polarization and social exclusion. In small rural communities this is very noticeable, especially in the east of Hungary. Such polarization leads to friction between the haves and the have-nots. The new focus on competition for employment following the regime change has broken up former community relationships based on working for the same cooperative farm and is often most strongly felt by women, as was also found in rural areas of the former GDR (van Hoven-Iganski, 2000). Many entrepreneurs feel that they are envied by their neighbours but others, especially women, have achieved social inclusion through participation in community organizations. Feelings of well-being, autonomy and having a choice are often developed through a successful business and can be the basis for social inclusion as well as building social capital (Oughton et al., 2003).

It is clear from our surveys that the decision to become an entrepreneur is embedded in the social and economic relations of the household and their family histories. An entrepreneurial family background may encourage the decision to become self-employed. But choices made by one member of the household may impose costs on another, as when women give up jobs they enjoy to work in their husband's business. On the other hand, some members of the family may look for non-material benefits rather than income, such as being able to stay at home with children or elderly parents and to give up commuting to work. Women often report that they miss the companionship of the workplace. However, running a business from home allows for flexibility in combining productive and reproductive tasks and a reduction of travel time. It also means no escape for women from housework and this is often seen as a benefit to the family by men.

Geopolitical Ruptures and Mindscapes of Difference

In our study we have also noticed the changes seen by Wastl-Walter and Váradi (2004), in terms of geopolitical changes and local perceptions on the eastern and western frontiers of Hungary. The political regime change, accompanied by privatization and the imposition of the rules of the market economy, has led to enormous social ruptures, especially in the borderlands. Globalization allows for flows of people and goods but in so doing brings new friction caused by the presence of foreigners. This is particularly felt in a country like Hungary which had previously been ethnically very homogeneous. The notion of connections, or open and bounded spaces, is integrally interwoven with concepts of borders but they may be seen as threat or opportunity. Mobility itself, in a formerly closed area, brings new possibilities and challenges. Concepts of how space may be open or closed have been heightened by these transformations. In Bulgaria it has been argued that the new 'open' society had made villages more closed because of loss of employment (Kozhuharova, 1997). The spatial pattern of uneven development in Hungary has been restructured by these border changes with proximity to western opportunities becoming valorized, especially in local mindscapes.

Western borders throughout Central and Eastern Europe are defined as zones of opportunity and as winners in the process of transition while eastern rural areas are stable and unchanging (Fassmann, 2000). The mindscapes of people living in the eastern and western borderlands are becoming increasingly differentiated. In the west the borderlands are seen as offering opportunity, whether through working or studying in Austria or providing goods and services for Austrian and other foreign consumers. In the east the border area is seen as isolated and threatened by flows of illegal immigrants. The optimism found on the western border is very different from the pessimism found in the villages in the eastern border area. Interestingly, a similar picture of optimism is found in villages on the western frontier of Romania, only a couple of miles across the border from eastern Hungary (Lelea, 2000). It appears that looking across a frontier at a more prosperous locale promotes mindscapes of opportunity and hope, whatever the reality. For western Romanians, as for western Hungarians, crossing the frontier provides opportunities for higher paying, though often illegal jobs. Language is less of a barrier in eastern Hungary where there are many ethnic Romanian Hungarians, than in western Hungary, although efforts to learn German are being encouraged both within families and in communities through educational exchanges for children and a focus on language teaching in village schools in the west. However, in both border areas the itinerant Romanian traders are disliked. Thus for both borders there is a growing regional transnational identity but also a feeling of regions in flux where there is the threat and the possibility of challenging local norms. Reactions to the local impacts of global change differ markedly on the two borders.

In particular, in the east, the memories of old borders seem to be stronger than in the west. Border changes have been more recent in the east and there is still a strong feeling of loss and anger at those changes which led to these areas being cut off from their previous urban centres and so to their current isolation. In the west, Burgenland belonged to the Hungarian part of the Austro-Hungarian Empire and was poorer than the area now in Hungary. The border changes following the break up of the Austro-Hungarian empire cut off Burgenland from its hinterland but these changes have been in place for so long that the previous knowledge of each other's languages has been lost and Austrian Burgenland is now seen as more prosperous than rural western Hungary. History plays little role in the mindscapes of people in western Hungary and there is often little interest in contact between villages on opposite sides of the border (Wastl-Walter and Váradi, 2002). These villages must now accept that

Bibliography

Adamik, Maria (1993), 'Feminism and Hungary', in Nanette Funk and Magda Mueller (eds), *Gender Politics and Post-Communist Reflections from Eastern Europe and the Former Soviet Union*, London: Routledge, pp. 203–12.

Agency for Regional Economic Development, Arad County (ADAR) (1996), *Regiunea economica Vest: Concept de Dezvoltare Regionala*, Timasoara: ARAD, Editura Orizonturi Universitare.

Agócs, P. and S. Agócs (1993), 'Entrepreneurship in Post-Communist Hungary', *The Journal of Social, Political and Economic Studies*, **18** (2), 159–79.

Andorka, Rudolph (1993), 'Rural-urban differences in income level and in living conditions in Hungary', *Landscape and Urban Planning*, **27**, 217–22.

Appadurai, A. (1996), *Modernity at large: Cultural dimensions of globalization*, Minneapolis, MN and London: University of Minnesota Press.

Bachtler, J. (1992), 'Regional Problems and Policies in Central and Eastern Europe', *Regional Studies*, **26** (7), 665–76.

Bakic-Hayden, M. (1995), 'Nesting orientalisms: the case of former Yugoslavia', *Slavic Review*, **54** (4), 917–31.

Balcsók, I. and Dancs, L. (2003), 'Munkaerő-piaci kapcsolatok az Északkelet-Alföldön, különös tekintettel a magyar-ukrá határ mentére' ('Labour market relations in the North-East of the Great Hungarian Plain with special attention to the Hungarian-Ukrainian Border Zone'), *Alföldi Tanulmányok*, **XIX**, 51–65.

Barclays Bank (1996), *Hungary*, Barclays Country Report, Poole, Dorset: Barclays Economic Department.

Barr, Nicholas (ed.) (1994), *Labor Markets and Social Policy in Central and Eastern Europe. The Transition and Beyond*, New York: Oxford University Press.

Barta, Barbara, András Klinger, Károly, Miltény and György Vukovich (1984), 'Fertility, female employment and policy measures in Hungary', *Women, Work and Development*, **6**, 1–19.

Bartlett, Will and Paul Hoggett (1996), 'Small firms in South East Europe: the importance of initial conditions', in Horst Brezinski and Michael Fritsch (eds), *The Economic Impact of New Firms in Post-Socialist Countries*, Cheltenham, UK and Brookfield, VT: Edward Elgar, pp. 151–74.

Beaverstock, J.V. and J.T. Boardwell (2000), 'Negotiating globalization, transnational corporations and global city financial centers in transient migration studies', *Applied Geography*, **20**, 227–304.

Bebbington, A. (2001), 'Globalized Andes? Livelihoods, landscapes and development', *Ecumene*, **8** (4), 369–80.

Bebbington, A and S. Batterbury (2001), 'Transnational livelihoods and landscapes: political ecologies of globalization', *Ecumene*, **8** (4), 369–80.

Benería, Lourdes (2003), *Gender, Development and Globalization. Economics as if All People Mattered*, London: Routledge.

Berényi, I. (1992), 'The Socio-economic Transformation and the Consequences of the Liberalisation of Borders in Hungary', in Z. Kovács (ed.), *New Perspectives in Hungarian Geography. A. Human Geographical Studies*, Budapest: Geographical Research Institute, Hungarian Academy of Sciences, pp. 45–58.

Berg, Nina Gunnerud (1997), 'Gender, place and entrepreneurship', *Entrepreneurship & Regional Development*, **9**, 259–68.

Blaskó, Zs. (2005), Nők, férfiak – keresőmunka, házimunka. Társadalmi Nem-és Kultúrakutató Központ, Corvinus Egyetem. Február 17. (accessed on http://demografia.hu/prgs/issp2002).

Bock, Bettina (2004a), 'It still matters where you live: Rural women's employment throughout Europe', in K. Hoggart and H. Buller (eds), *Women in the European Countryside*, London: Ashgate, pp. 69–89.

Bock, Bettina B. (2004b), 'Fitting in and multi-tasking: Dutch Farm Women's Strategies in Rural Entrepreneurship', *Rural Sociology* **44** (3), 245–60.

Bogyó, István (1995), 'Moving toward Sustainability: Case studies from the Debrecen SBDC', in D.S. Fogel, M.E. Harrison and F. Hoy (eds), *Moving to Sustainability: How to keep small business development centers alive*, Aldershot: Avebury, pp. 91–98.

Böhm, Antal (1995), 'Local Politics in Border Regions in Central Europe', *The Annals of the American Academy of Politics and Social Science*, **540**, 137–44.

Bollobás, Enikö (1993), '"Totalitarian Lib": the Legacy of Communism for Hungarian Women', in N. Funk and M. Mueller (eds), *Op. cit.*, pp. 201–6.

Böröcz, József (1993), 'Simulating the great transformation: property change under prolonged informality in Hungary', *Archives of European Sociology*, **XXXIV**, 81–107.

Brah, A. (1996), *Cartographies of diaspora. Contesting identities*, London and New York: Routledge.

Brettell, C. (2003), *Anthropology and migration: Essays on transnationalism, ethnicity and identity*, Walnut Creek, Lanham, New York and Oxford: Altamira Press.

Brezinski, Horst and Michael Fritsch (eds) (1996), *The economic impact of new firms in Post-socialist countries. Bottom-up transformation in Eastern Europe*, Cheltenham, UK and Brookfield, USA: Edward Elgar.

Burant, Stephen R. (1990), *Hungary, a country study*, Washington, DC: Library of Congress.

Burt, R.S. (1998), 'The Gender of Social Capital', *Rationality and Society*, **10** (1), 5–46.

Carter, Sara (1993), 'Female business ownership: current research and possibilities for the future', in Sheila Allen and Carole Truman (eds), *Women in Business. Perspectives on women entrepreneurs*, London and New York: Routledge, pp. 148–60.

CCET (The Centre for Cooperation with Economies in Transition) (1995a), *Social and Labour Market Policies in Hungary*, Paris: OECD.

CCET (The Centre for Cooperation with Economies in Transition) (1995b), *The Regional dimension of unemployment in Transition countries: A challenge for labour market and social policies*, Paris: OECD.

Ciechocińska, Maria (1993), 'Gender Aspects of Dismantling the Command Economy in Eastern Europe: the Polish Case', *Geoforum*, **24** (1), 31–44.

Clement, Norris G. (2004), 'Economic forces shaping the Borderlands', in Morehouse et al. (eds), *Op. cit.*, pp. 41–62.

Corrin, Chris (ed.) (1992a), *Superwoman and the Double Burden. Women's Experience of Change in Central and Eastern Europe and the former Soviet Union*, Toronto: Second Story Press.

Corrin, Chris (1992b), 'Gendered Identities. Women's Experience of Change in Hungary', in Shrin Rai, Hilary Pilkington and Annie Phizacklea (eds), *Women in the Face of Change. The Soviet Union, Eastern Europe and China*, London: Routledge, pp. 167–85.

Corrin, Chris (1994), *Magyar women: Hungarian women's lives, 1960s–1990s*, Basingstoke, Hants: The Macmillan Press; New York: St Martin's Press.

Conway, D. and J. Cohen (1998), 'Consequences of migration and remittances for Mexican transnational communities', *Economic Geography*, **74** (1), 26–44.

Crang, P., C. Dwyer and P. Jackson (2003), 'Transnationalism and the spaces of commodity culture', *Progress in Human Geography*, **27**, 438–56.

Csaki, Csaba and Zvi Lerman (eds) (2000), *Structural Change in the Farming Sectors in Central and Eastern Europe*, World Bank Technical Paper No. 465, Washington, DC: World Bank.

Csefalvay, Z. (1994), 'The Regional Differentiation of the Hungarian Economy in Transition', *GeoJournal*, **32** (4), 351–61.

Csete L. (1995), 'A magyarországi állami gazdaságok privatizálása és átalakítása' ('Privatisation and the process of change in state farms in Hungary'), *Gazdálkodás*, **XXXIX** (5), 20–35.

Csite A. and I. Kovách (1995), 'Poszt-szocialista átalakulás Közép- és Kelet-Európa rurális társadalmaiban' ('The post-socialist transition of rural society in Central and Eastern Europe'), *Szociológiai Szemle*, **2**, 49–72.

Csordás, László and Géza Szabó (1993), 'Rural tourism: Its organisation and development in the Alföld', *Alföldi Tanulmányok*, **15**, 142–61.

Curran, James and Robert Blackburn (1991), *Path of enterprise: the future of small business*, London: Routledge.

Denich, Bette (1990), 'Paradoxes of Gender and Policy in Eastern Europe: A Discussant's Comments', *East European Quarterly*, **XXIII** (4), 499–506.

Dumont, René (1957), *Types of Rural Economy. Studies in World Agriculture*, London: Methuen & Co.

Eberhardt, Eva (1991), *Women of Hungary*, Brussels: Commission of the European Communities Directorate-General Information, Communication, Culture: Women's Information Service.

Economist, The (1996a), *Pocket World in Figures*, London: Profile Books Ltd.

Economist, The (1996b), 'Hungarian Retailing: Magyar malls', *The Economist*, 9 November, 83.

Economist, The (2002), *Pocket World in Figures*, London: Profile Books Ltd.

Economist, The (2005), *Pocket World in Figures*, London: Profile Books Ltd.

Einhorn, Barbara (1993a), *Cinderella goes to Market. Citizenship, Gender and Women's Movements in East Central Europe*, New York and London: Verso.

Einhorn, Barbara (1993b), *The Impact of the Transition from Centrally Planned Economies on Women's Employment in East Central Europe*, Bridge, Report No. 18, Geneva: ILO.

Erdei, Ferenc and Antal Vermes (1995), 'The struggle for balance in rural Hungary', in D.E. Kideckel (ed.), *East European communities: the struggle for balance in turbulent times*, Boulder, CO: Westview Press, pp. 101–14.

Enyedi, Gy. (1990), *New Basis for Regional and Urban Policies in East-Central Europe*, Discussion Paper No. 9, Pécs, Hungary: Centre for Regional Studies of the Hungarian Academy of Sciences.

Enyedi, G. (1994), 'Regional and Urban Development in Hungary until 2005', in Z. Hajdú and G. Horváth (eds), *European Challenges and Hungarian Responses in Regional Policy*, Pécs: Centre for Regional Studies of the Hungarian Academy of Sciences, pp. 239–53.

Fajth, Gaspar and Judit Lakatos (1994), 'Unemployment in Hungary', in T. Boeri (ed.), *Unemployment in Transition Countries: Transient or Persistent*, Paris: OECD, pp. 169–95.

Falussy B. and Gy. Vukovich (1996), 'Az idő mérlegén (1963–1993)', in R. Andorka, T. Kolosi and Gy. Vukovics (eds), *Társadalmi Riport 1996* (Social Report 1996), Budapest: TÁRKI–Századvég, pp. 70–103.

Fassmann, Heinz (2000), 'Regions in upheaval. Conceptual framework and empirical findings of the regional transformation research', in G. Horváth (ed.), *Regions and Cities in the Global World*, Pécs, Hungary: Centre for Regional Studies, pp. 126–40.

Feragó, László (1999), 'Regional "Winners" and "Losers"', in Z. Hajdú (ed.), *Regional Processes and Spatial Structures in Hungary in the 1990s*, Pécs, Hungary: Centre for Regional Studies of the Hungarian Academy of Sciences, pp. 316–27.

Ferge, Zsuzsa (1992), 'Unemployment in Hungary: the Need for a New Ideology', in B. Deacon (ed.), *Social Policy, Social Justice and Citizenship in Eastern Europe*, Aldershot: Avebury, pp. 158–75.

Ferrão, João and Raul Lopes (2004), 'Understanding Peripheral Rural Areas as Contexts for Economic Development', in Lois Labrianidis (ed.), *The Future of Europe's Rural Peripheries*, Aldershot: Ashgate, pp. 31–61.

Fodor, Eva (1994), 'The Political Woman? Women in Politics in Hungary', in M. Rueschmeyer (ed.), *Op. cit.*, pp. 171–99.

Fodor, Istsván (1999), 'The effect of the socio-economic transition on the Hungarian environment', in Z. Hajdú (ed.), *Regional Processes and Spatial Structures in Hungary in the 1990s*, Pécs, Hungary: Centre for Regional Studies of the Hungarian Academy of Sciences, pp. 328–54.

Fretwell, David and Richard Jackman (1994), 'Labor Markets: Unemployment', in N. Barr (ed.), *Labor Markets and Social Policy in Central and Eastern Europe*, published for the World Bank and the London School of Economics and Political Science, Washington, DC and New York: Oxford University Press.

Foster, J.B. and W.J. Froman (eds) (2002), *Thresholds of western culture: Identity, postcoloniality, transnationalism*, New York and London: Continuum.

Funk, Nanette and Magda Mueller (eds) (1993), *Gender Politics and Post-Communism. Reflections from Eastern Europe and the former Soviet Union*, London and New York: Routledge.

Gábor, István R. (1997), 'Too many, too small: Small entrepreneurship in Hungary – Ailing or Prospering', in G. Grabner and D. Stark (eds), *Restructuring Networks in Post-Socialism: Legacies, Linkages and Localities*, Oxford: Oxford University Press, pp. 158–75.

Gal, Susan (1994), 'Gender in the Post-Socialist Transition: The abortion debate in Hungary', *East European Politics and Societies*, **8** (2), 256–86.

Geographical Research Institute, Hungarian Academy of Sciences (1994), *National Atlas of Hungary, Supplementary Map Lift-Out Series Part Two, Population and Demographic Trends, 1980–1989*, Budapest: Hungarian Academy of Sciences.

Geographical Research Institute, Hungarian Academy of Sciences (1995), *National Atlas of Hungary, Supplementary Map Lift-Out Series Part Three, International Migration, 1980–1992*, Budapest: Hungarian Academy of Sciences.

Gere, Ilona (1996), 'Women Entrepreneurs in Today's Hungarian Society', in Z. Laczko and A. Soltesz (eds), *The Status of Women in the Labour Market in Hungary – Entrepreneurship as Alternative*, Budapest: SEED Foundation, pp. 31–42.

Gömöri, E. (1980), 'Hungary', in *Work and Family Life: the role of the social infrastructure in Eastern European countries*, Geneva: International Labour Office, pp. 28–41.

Goven, Joanna (1993), 'Gender Politics in Hungary: Autonomy and antifeminism', in Nanette Funk and Magda Mueller (eds), *Gender Politics and Post-communism: Reflections from Eastern Europe and the former Soviet Union*, New York: Routledge.

Goverde, Henri, Henk de Haan and Mireia Baylina (eds) (2004), *Power and Gender in European Rural Development*, Aldershot: Ashgate.

Groen, R. and A.Visser (1993), *Development Chances for Békés County (A Terület-Fejlesztés Esélyei Békés Megyében)*, Utrecht and Békéscsaba: Faculty of Geographical Sciences of the Utrecht University and Centre for Regional Studies of the Hungarian Academy of Sciences, Alföld Institute, Békéscsaba.

Guarnizo, L.E. (2003), 'The economics of transnational living', *The International Migration Review*, **37** (3), 666–99.

Habuda, Judit (1995), 'Post-socialist transformation in Hungary: Entering a second stage?', in J. Hausner, B. Jessop and K. Nielsen (eds), *Strategic Choice and Path-dependency in Post-Socialism: Institutional Dynamics in the Transformation Process*, Brookfield, VT and Aldershot: Edward Elgar, pp. 309–24.

Hajdù, Zoltán (1996), 'Emerging Conflict or Deepening Cooperation? The Case of the Hungarian Border Regions', in James Scott, Alan Sweedler, Paul Ganster and Wolf-Dieter Eberwein (eds), *Border Regions in Functional Transition. European and North American Perspectives on Transboundary Interaction*, Berlin: Institute for Regional Development and Structural Planning, Regio Series of the IRS No. 9, pp. 139–52.

Hamilton, F.E.I. (1999), 'Transformation and space in Central and Eastern Europe', *The Geographical Journal*, **165** (2), 135–44.

Haney, Lynne (1994), 'From proud worker to good mother: Women, the State and Regime Change in Hungary', *Frontiers*, **XIV** (3), 113–50.

Hantó, Zs. and A. Oberschall (1997), 'A piacgazdaság születése: A magyar mezőgazdaság a szocializmus után' ('Birth of the Market Economy: Hungarian agriculture after the socialist regime'), *Gazdálkodás*, **XLI** (6), 33–45.

Hardi, T. (1999), 'A határ és az ember-Az osztrák – magyar határ mentén élők képe a határról és a "másik oldalról"' ('The border and man – a picture of the people living on the Austrian-Hungarian Border, on the Border and on the "other side"'), in M. Nárai and J. Rechnitzer (eds), *Elválaszt és összeköt a határ*, Pécs and Győr: Regional Research Centre, Hungarian Academy of Sciences, pp. 159–90.

Hardill, I. (1998), 'Trading Places: case studies of the labour market experiences of women in rural in-migrant households', *Local Economy*, **13**, 102–13.

Hárs, Ágnes, Gy. Kövári and G. Nagy (1991), 'Hungary faces unemployment', *International Labour Review*, **130** (2), 65–175.

Heather, Barbara, D. Lynn Skillen, Jennifer Young and Theresa Vladicka, 'Women's Gendered Identities and the Restructuring of Alberta', *Sociologia Ruralis*, **45** (1/2), 86–97.

Heinen, Jacqueline (1994), 'The re-integration into work of unemployed women: Issues and policies', in T. Boeri (ed.), *Op. cit.*, pp. 311–33.

Herald Tribune (2004), 'The opening of Reconciliation Park', 26 April, p. 3.

Hisrich, Robert D. and Gyula Fülöp (1995), 'Hungarian entrepreneurs and their enterprises', *Journal of Small Business Management*, **33** (3), 88–94.

Hisrich, Robert D. and Gyula Fülöp (1997), 'Women entrepreneurs in family business: The Hungarian Case', *Family Business Review*, **10** (3), 281–302.

Hitchcock, P. (2003), *Imaginary states: Studies in cultural transnationalism*, Urbana and Chicago: University of Illinois Press.

Horváth, G. (1999a), 'Changing Hungarian Regional Policy and Accession to the European Union', *European Urban and Regional Studies*, **6** (2), 166–77.

Horváth, G. (1999b), 'Regional effects of the transition in East Central Europe', in Z. Hajdú (ed.), *Regional Processes and Spatial Structures in Hungary in the 1990s*, Pécs, Hungary: Centre for Regional Studies of the Hungarian Academy of Sciences, pp. 9–33.

Hrubos, Ildikó (1994), 'Women in the Labour Market in Hungary', *Women's Studies International Forum*, **17** (2/3), 311–12.

Hudson, A. (2001), 'NGOs, transnational advocacy networks: from "legitimacy" to "political responsibility"?', *Global Networks: A Journal of Transnational Affairs*, **1** (4), 331–52.

Hudson, Ray (2003), 'European Integration and New Forms of Uneven Development: But Not the End of Territorially Distinctive Capitalisms in Europe', *European Urban and Regional Studies*, **10** (1), 49–67.

Hungarian Central Statistical Office (1997), *Regional Statistical Yearbook, 1996*, Budapest: Statistical Office.

Hungarian Central Statistical Office (1997), *Statistical Yearbook for Békés County, 1996*, Békés County Management, Budapest: Statistical Office.

Hungarian Central Statistical Office (1997), *Statistical Yearbook for Győr-Moson-Sopron County, 1996*, Győr-Moson-Sopron County Management, Budapest: Statistical Office.

Hungarian Central Statistical Office (2003), *Regional Statistical Yearbook, 2002*, Budapest: Statistical Office.

Hungarian Central Statistical Office (2003), *Statistical Yearbook for Békés County, 2002*, Békés County Management, Budapest: Statistical Office.

Hungarian Central Statistical Office (2004), *Statistical Yearbook for Győr-Moson-Sopron County, 2003*, Győr-Moson-Sopron County Management, Budapest: Statistical Office.

Huseby-Darvas, Eva V. (1990), 'Migration and gender: Perspectives for rural Hungary', *East European Quarterly*, **XXIII** (4), 487–98.

Iganski, Bettina (1999), 'The Meaning of Women's "Second Family" for Current Patterns of Discontinuity in Rural East Germany', in P. Cooke and J. Grix (eds), *East Germany Continuity and Change*, Amsterdam and Atlanta, GA: Rodopi, pp. 151–61.

(Van-Hoven)-Iganski, Bettina (2000), *Made in the GDR. The Changing Geographies of Women in the Post-Socialist Rural Society in Mecklenburg-Westpommerania*, Netherlands Geographical Studies 267, Utrecht/Groningen: Koningklijk Nederlands Aardsdrijkskundig Genootschap/Faculteit der Ruimtelijke Wetenschappen.

Ingham, Mike and Keith Grime (1994), 'Regional unemployment in Central and Eastern Europe', *Regional Studies*, **28** (8), 811–17.

International Labour Office (ILO) (2003), *Yearbook of Labour Statistics*, Geneva: ILO.

Jánossomorja (1998), Interview with mayor, 25 June.

Jarvis, Susan J. and John Mickelwright (1992), *The Targeting of Family Allowances in Hungary*, EUI Working paper ECO No. 9296, San Domenico, Italy: European University Institute.

Johnston, R.J., Derek Gregory, Geraldine Pratt and Michael Watts (2000), *The Dictionary of Human Geography*, Fourth Edition, Malden, MA and Oxford: Blackwell.

Kaizer, Alenka (1995), 'The real-wage-employment relationship and unemployment in transition economies: the case of Slovenia and Hungary', *Eastern European Economics*, **33** (4), 55–78.

Kalantaridis, Christos (2004), 'Entrepreneurial Behaviour in Rural Contexts', in L. Labrianidis (ed.), *The Future of Europe's Rural Peripheries*, Aldershot: Ashgate, pp. 62–85.

Karalyos Zs. (1997), 'A szövetkezeti termőföld-használat időszerű kérdései Hajdú-Bihar megyében' ('Questions of agricultural land-use in Hajdú-Bihar County'), *Gazdálkodás*, **XLI** (4), 72–76.

Kelly, Rita Mae, Jane H. Bayes, Mary Hawkesworth and Brigitte Young (eds) (2001), *Gender, Globalization, and Democratization*, Lanham, Boulder, New York, Oxford: Rowman and Littlefield Publishers Inc.

Kiss, Judit (1991), 'The Second No: Women in Hungary', *Feminist Review*, **39**, 49–57.

Kiss, Judit (2003), 'A Magyar mezőgazdaság világgazdasági mozgástere' ('The position of Hungarian agriculture in the world economy'), Budapest: Akadémiai Kiadó, p. 17.

Kiss, S. (1999), Personal communication by the mayor of Iratoşu, Variaşu Mar and Variaşu Mic, August.

Köllô, János (1995), 'Unemployment and the prospects for employment policy in Hungary', in M. Jackson, J. Koltay and W. Beisbrouck (eds), *Unemployment and evolving labour markets in Central and Eastern Europe*, Aldershot: Avebury, pp. 183–227.

Koncz, Katalin (1995), 'The position of Hungarian women in the process of regime change', in B. Lobodzińska (ed.), *Family, Women and Employment in Central-Eastern Europe*, Westport, CT: Greenwood Press, pp. 131–48.

Koncz, Katalin (2000), 'Transitional Period and Labor Market Characteristics in Hungary', in Marnia Lazreg (ed.), *Making the Transition Work for Women in Europe and Central Asia*, World Bank Discussion Paper No. 411, Washington, DC: The World Bank, pp. 26–41.

Kong, L. (1999), 'Globalization and Singaporean transmigration: re-imagining and negotiating national identity', *Political Geography*, **18**, 563–89.

Kovach, Imre (1991), 'Rediscovering small-scale enterprise in rural Hungary', in S. Whatmore, P. Lowe and T. Marsden (eds), *Rural Enterprise: Shifting Perspectives on Small-Scale Production*, London: David Fulton Publishers, pp. 78–96.

Kovács, K. (1996), 'The Transition in Hungarian Agriculture 1990–1993. General Tendencies, Background Factors and the Case of the "Golden Age"', in R. Abrahams (ed.), *After Socialism. Land Reform and Social Change in Eastern Europe*, Providence, RI and Oxford: Berghahn Books, pp. 51–84.

Kovács, Katalin and Mónika Váradi (2000), 'Women's Life Trajectories and Class Formation in Hungary', in S. Gal and G. Kligman (eds), *Reproducing Gender, Politics, Publics, and Everyday Life after Socialism*, Princeton, NJ: Princeton University Press, pp. 176–99.

Kovács, Teréz (1993), *Borderland Situation as seen by a Sociologist*, Discussion Paper No. 18, Pécs, Hungary: Centre for Regional Studies of the Hungarian Academy of Sciences.

Kovács, Teréz (1999), 'Regional Disparities in the Privatisation of Land', in Z. Hajdú (ed.), *Regional Processes and Spatial Structures in Hungary in the 1990s*, Pécs, Hungary: Centre for Regional Studies of the Hungarian Academy of Sciences, pp. 99–122.

Kovács, Zoltán (1989), 'Border changes and their effect on the structure of Hungarian society', *Political Geography Quarterly*, **8** (1), 79–86.

Kozhuharova, Veska (ed.) (1997), *The Bulgarian Village and Globalisation Processes*, Sofia: Bulgaria Rusticana.

Kukorelli, Sz. Irén (1997), 'Economic and social features and trends of small regions in North Transdanubia', *Tér és Társadalom* (*Space and Society*), **1**, 147–82.

Kukorelli, Sz. I., L. Dancs, Z. Hajdu, J. Kugler and I. Nagy (2000), 'Hungary's Seven Border Regions', *Journal of Borderlands Studies – Special Number – European Perspectives on Borderlands*, Joachim Blatter and Norris Clement (eds), **XV**, (1), Spring, 1–32.

Kulcsar, Laszlo and David Brown (2000), 'Rural families and rural development in Central and Eastern Europe', in D. Brown and A. Bandlerová (eds), *Rural Development in Central and Eastern Europe*, Nitra, Slovakia: Slovak Agricultural University, pp. 111–17.

Kulcsár, Rózsa (1985), 'The socio-economic conditions of women in Hungary', in S.L. Wolchik and A.G. Meyer (eds), *Women, State and Party in Eastern Europe*, Durham, NC: Duke University Press, pp. 195–213.

Kürti, László (1990), '"Red Csepel": Working Youth in a Socialist Firm', *East European Quarterly*, **XXIII** (4), 445–68.

Kuus, Merje (2004), 'Europe's eastern expansion and the re-inscription of otherness in East-Central Europe', *Progress in Human Geography*, **28** (4), 472–89.

Kwiecińska-Zdrenka, M. (2001), 'The Polish and Hungarian Countryside a Decade after the System Transformation. Review of *Rural societies under Communism and beyond. Hungarian and Polish Perspectives*, edited by Pavel Starosta, Imre Kovach and Krzysztof Gorlach, Lodz: Lodz University Press, 1999, *Eastern European Countryside*, **7**, 127–32.

Labrianidis, Lois (2004), 'Introduction', in Lois Labrianidis (ed.), *The Future of Europe's Rural Peripheries*, Aldershot: Ashgate, pp. 31–61.

Laczló, Zsusa (1996), 'Helping businesses in the "subsistence economy" in Hungary', in J. Levitsky (ed.), *Small Business in Transition Economies: Promoting Enterprise in Central and Eastern Europe and the former Soviet Union*, London: IT Publications, pp. 30–33.

Laczó, F. (1994), 'A tulajdonszerkezet változása a mezőgazdaságban' ('The change of the ownership-structure in agriculture'), *Gazdálkodás*, **XXXVIII** (3), 1–19.

Lampland, Martha (1990), 'Unthinkable Subjects: Women and Labor in Socialist Hungary', *East European Quarterly*, **XXIII**, (4), 389–98.

Lampland, Martha (1995), *The Object of Labor. Commodification in Socialist Hungary*, Chicago and London: University of Chicago Press.

Laschewski, L. and R. Siebert (2003), 'Social Capital Formation in Rural East Germany', in H. Goverde, H. de Haan and M. Baylina (eds), *Power and Gender in European Rural Development*, Aldershot, UK: Ashgate, pp. 20–31.

Lazreg, Marnia (ed.) (2000), *Making the Transition Work for Women in Europe and Central Asia. World Bank Discussion Paper No. 411*, Washington, DC: The World Bank.

Lelea, Margareta (2000), 'Women's Entrepreneurship in Rural Romania Bordering Hungary', unpublished Honors thesis, University of California, Davis.

Lengyel, G. and I.J. Tóth (1994), *The Spread of Entrepreneurial Inclinations in Hungary*, Studies in Public Policy No. 224, Glasgow: Centre for the Study of Public Policy, University of Strathclyde.

Le, Thuy-Doan (2005), 'Breaking with Tradition', *Sacramento Bee*, 20 March, p. D.1.

Lichtenberger, Elisabeth (2000), 'The globalization of the economy and the effects of EU policy: the case of Austria', in G. Horváth (ed.), *Regions and Cities in the Global World*, Pécs, Hungary: Centre for Regional Studies of the Hungarian Academy of Sciences, pp. 115–25.

Lobodzińska, Barbara (ed.) (1995), *Family, women and employment in Central-Eastern Europe*, Westport, CT: Greenwood Press.

Lovászi, Cs. (1999), 'Termőföld-tulajdonváltás Magyarországon 1988–1998' ('The change of agricultural-land-owners in Hungary'), *Számadás a Talentumról sorozat* (*Series: The rendering of the account*), Budapest: Állami Privatizációs és Vagyonkezelő Rt.

Maltby, Tony (1994), *Women and Pensions in Britain and Hungary: A cross-national and comparative case study of social dependency*, Aldershot: Avebury.

Makara, Klára (1992), 'A woman's place', *New Hungarian Quarterly*, **33** (126), 93–105.

Matynia, Elzbieta (1995), 'Finding a Voice: Women in Postcommunist Central Europe', in A. Basu (ed.), *The Challenge of Global Feminisms: Women's Movements in Global Perspective*, Boulder, CO: Westview Press.

Maurel, Marie-Claude (2000), 'Patterns of post-socialist transformation in the rural areas of Central Europe', in G. Horváth (ed.), *Regions and Cities in the Global World*, Pécs, Hungary: Centre for Regional Studies, pp. 141–58.

McIntyre, Robert J. (1985), 'Demographic policy and sexual equality: Value conflicts and policy appraisal in Hungary and Romania', in S.L. Wolchik and A.G. Meyer (eds), *Women, State and Party in Eastern Europe*, Durham, NC: Duke University Press, pp. 195–213.

Meinhof, Ulrike H. (ed.) (2002), *Living (with) Borders. Identity Discourses on East-West Borders in Europe*, London: Ashgate.

Meinhof, Ulrike H., Heidi Armbruster and Craig Rollo (2002), 'Identity Discourses on East-West Borders in Europe: An Introduction', in Ulrike H. Meinhof (ed.), *Living (with) Borders. Identity Discourses on East-West Borders in Europe*, London: Ashgate, pp. 1–13.

Micklewright, John and Gyula Nagy (1996), 'Labour market policy and the unemployed in Hungary', *European Economic Review*, **40** (3–5), 819–28.

Mieczkowski, Bogdan (1985), 'Social services for women and childcare facilities in Eastern Europe', in Wolchik and Mayer (eds), *Op. cit.*, pp. 257–69.

Mihályi, P. (1998), 'A kárpótlás' ('The compensation process'), *Számadás a Talentumról sorozat (Series: The rendering of the account)*, Budapest: Állami Privatizációs és Vagyonkezelő Rt.–Kulturtrade Kiadó Kft.

Miles, Miranda and Jonathan Crush (1993), 'Personal Narratives as Interactive Texts: Collecting and Interpreting Migrant Life-Histories', *Professional Geographer*, **45** (1), 84–96.

Minniti, M. and P. Arenius (2003), *Women in Entrepreneurship. The Entrepreneurial Advantage of Nations: First Annual Global Entrepreneurship Symposium*, New York: United Nations.

Mitchell, K. (2002), 'Cultural geographies of transnationality', in K. Anderson, M. Domosh, S. Pile and N. Thrift (eds), *The Handbook of Cultural Geography*, London: Sage Publications, pp. 74–87.

Moghadam, Valentine M. (ed.) (1992), *Privatisation and Democratization in Central and Eastern Europe and the Soviet Union: The Gender Dimension*, Helsinki: World Institute for Development Economics Research of the United Nations University.

Momsen, Janet H. (2000), 'Spatial transformations and economic restructuring in post-socialist Hungary', in G. Horváth (ed.), *Regions and Cities in the Global World*, Pécs, Hungary: Centre for Regional Studies, pp. 202–19.

Momsen, Janet H. (2002), 'Gender and Entrepreneurship in Post-Communist Hungary', in A. Rainnie, A. Smith and A. Swain (eds), *Work, Employment and Transition in Post-Communist Europe*, London: Routledge, pp. 155–69.

Momsen, Janet H. (2004), *Gender and Development*, London and New York: Routledge.

Momsen, Janet H. and Anne Jervell (2001), 'Rural tourism in California and Norway', paper presented at European Rural Sociology Society conference in Dijon, France, 5–9 September.

Momsen, Janet H. and Irén K. Szörényi (2002), 'Gender and Rural Entrepreneurship in Post-communist East and West Hungary', *Eastern European Countryside*, October, 109–18.

Morehouse, Barbara J. (2004), 'Theoretical Approaches to Border Spaces and Identities', in V. Pavlakovich-Kochi, B.J. Morehouse and D. Wastl-Walter (eds), *Challenged Borderlands: Transcending Political and Cultural Boundaries*, Aldershot: Ashgate, pp. 19–40.

Morrell, I.A. (1999a), *Emancipation's Dead-End Road? Studies in the Formation and Development of the Hungarian Model for Agriculture and Gender (1956–1989)*, Uppsala: Acta Universitatis Upsaliensis, Studia Sociologica Upsaliensia 46.

Morrell, Ildikó Asztalos (1999), 'Post-socialist rural transformation and gender construction processes', paper presented at the conference on *Gender and Rural Transformations in Europe* at Wageningen Agricultural University, Wageningen, the Netherlands, 14–17 October.

Moss, Pamela (ed.) (2001), *Feminist Geography in Practice*, Oxford: Blackwell.

Muller, E., A. Zenker and T. Dőry (2002), 'Regional innovation capacities and economic transition: the example of West Transdanubia', in O. Pfirrmann and G.H. Walter (eds), *Small Firms and Entrepreneurship in Central and Eastern Europe*, New York and Heidelberg: Physica-Verlag, pp. 251–71.

Nagy, B. (1996), 'Sikeres női vállalkozók a kilencvenes évek közepén', in Zs. Laczkó and A. Soltész (eds), *A nők munkaerőpiaci helyzete Magyarországon-a vállalkozás mint alternativa*, Budapest: SEED Kisvállalkozás-fejlesztési Alapitvány, pp. 59–65.

Nagy, B. (1997), 'New Career Perspectives – Women Entrepreneurs in Hungary', in F. Feischmidt, E. Magyari-Vincze and V. Zentai (eds), *Women and Men in East European Transition*, Cluj-Nopoca, Romania: Editura Fundaţiei Pentru 100-Studii Europenne, pp. 100–109.

Narayan, D. with R. Patel, K. Schafft, A. Rademacher and S. Koch-Schulte (2000), *Voices of the Poor. Can Anyone Hear Us?* New York: Oxford University Press for the World Bank.

Nemenyi, M. and A. Kende (1999), 'Anyák és lányok', *Replika*, **35**, 117–41.

Nyberg-Sørensen, N. (1998), 'Narrating across Dominican worlds', in M.P. Smith and L.E. Guarnizo (eds), *Transnationalism from Below*, New Brunswick, NJ: Transaction Publishers, pp. 241–69.

Oberhauser Ann (2004), 'Small business and gender in South Africa', paper presented at American Association of Geographers conference, Philadelphia, 14–20 March.

Orosz, Eva (1990), 'Regional inequalities in the Hungarian health system', *Geoforum*, **21** (2), 245–59.

Oughton, Elizabeth, Jane Wheelock and Susan Baines (2003), 'Micro-businesses and Social Inclusion in Rural Households: A Comparative Analysis', *Sociologia Ruralis*, **43** (4), 331–48.

Pál, Ágnes and Imre Nagy (2003), 'The economic relationships of the Hungarian-Romanian border zone', paper presented at the Hungarian Academy of Sciences, Regional Research Centre, conference on borders, Békéscsaba, 21–25 September.

Paul, Leo (1992), 'Developments in rural areas in Eastern Europe since the Second World War: An Overview', in Paulus Huigen, Leo Paul and Kees Volkers (eds), *The changing function and position of rural areas in Europe*, Utrecht: Koninkliijk Nederlands Aardrijkskundig Genootschap, pp. 101–108.

Paul, Leo, Imre Simon, Bálint Czatári, Csilla Keresztes Nagy, Rob Groen and Adriaan Visser (1992), 'Rural areas and rural policy in Hungary', in Paulus Huigen, Leo Paul and Kees Volkers (eds), *The changing function and position of rural areas in Europe*, Utrecht: Koninkliijk Nederlands Aardrijkskundig Genootschap, pp. 153–68.

Pavlakovich-Kochi, Vera, Barbara J. Morehouse and Doris Wastl-Walter (2004), *Challenged Borderlands: Transcending Political and Cultural Boundaries*, Aldershot: Ashgate.

Pongrácz, T.-né and S. Molnár (1996), 'Gyermeket nevelni' ('To raise a child'), in R. Andorka, T. Kolosi and Gy. Vukovics (eds), *Társadalmi Riport 1996 (Social Report 1996)*, Budapest: TÁRKI–Századvég, pp. 332–51.

Portes, A. (1996), 'Transnational communities: their emergence and significance in the contemporary world system', in R.P. Korzeniewicz and W.C. Smith (eds), *Latin America in the world economy*, Westport, CN: Greenwood Press, pp. 151–68.

Portes, A. (1998), 'Social Capital: Its Origins and Applications in Modern Sociology', *Annual Review of Sociology* **24**, 1–24.

Putnam, R.D. (1993), *Making Democracy Work*, Princeton, NJ: Princeton University Press.

Quack, S. and F. Maier (1994), 'From state socialism to market economy – women's employment in East Germany', *Environment and Planning A*, **26** (8), 1257–76.

Rechnitzer, János (2000), *The Features of the Transition of Hungary's Regional System, Discussion Paper No. 32*, Pécs: Centre for Regional Studies of the Hungarian Academy of Sciences.

Répássy, Helga (1991), 'Changing Gender Roles in Hungarian Agriculture', *Journal of Rural Studies*, **7** (1/2), 23–30.

Répássy, Helga and David Symes (1993), 'Perspectives on Agrarian Reform in East-Central Europe', *Sociologia Ruralis*, **XXXIII** (1), 81–91.

Róna-Tas, Ákos (1997), *The Great Surprise of the Small Transformation. The Demise of Communism and the Rise of the Private Sector in Hungary*, Ann Arbor: University of Michigan Press.

Róna-Tas, Ákos and Gy. Lengyel (1997), 'Entrepreneurs and Entrepreneurial Inclinations in Post-Communist East-Central Europe', *International Journal of Sociology*, **27** (3), 3–14.

Rueschmeyer, Marilyn (ed.) (1994), *Women in the Politics of Post-Communist Eastern Europe*, Armonk, NY and London: M.E. Sharpe.

Sen, Amartya (2000), *Development as freedom*, New York: Anchor Books.

Sik, E. (1999), 'Magyarok az osztrák munkaearőpiacon', *Századvég*, **12**, 60–85.

de Silva, Lalith (1993), 'Women's Emancipation under Communism. A Re-evaluation', *East European Quarterly*, **XXVII** (3), 301–15.

Sizoo, Edith (ed.) (1997), *Women's lifeworlds. Women's narratives on shaping their realities*, London and New York: Routledge.

Skuras, Dimitris, Nicolas Meccheri, Manuel Belo Moreira, Jordi Rosell and Sophia Stathopoulu (2005), 'Entrepreneurial human capital accumulation and the growth of rural businesses: a four-country survey in mountainous and lagging areas of the European Union', *Journal of Rural Studies*, **21** (1), 67–79.

Smith, A. (2000), 'Employment restructuring and household survival in "post-communist" transition: rethinking economic practices in Eastern Europe', *Environment and Planning A*, **32**, 1759–80.

Smith, A. and S. Ferenčíková (1998), 'Inward Investment, Regional Transformations and Uneven Development in Eastern and Central Europe', Enterprise Case Studies from Slovakia. *European Urban and Regional Studies*, **5** (2), 155–73.

Smith, A. and J. Pickles (1998), *Theorising Transition: the political economy of transition in post-communist countries*, London and New York: Routledge.

Smith, A. and A. Swain (1998), 'Regulating and institutionalizing capitalisms. The micro-foundations of transformation in Eastern and central Europe', in J. Pickles and A. Smith (eds), *The Political Economy of Post-Communist Transformations*, London: Routledge, 25–53.

Smith, N. (2000), 'Uneven development', in Johnston et al. (eds), *Op. cit.*, 867–69.

Somogyi, Gabriella Kraft (1999), 'The role of tourism in Regional Development', in Z. Hajdú (ed.), *Regional Processes and Spatial Structures in Hungary in the 1990s*, Pécs, Hungary: Centre for Regional Studies of the Hungarian Academy of Sciences, pp. 156–79.

Staeheli, Lynn A. and Victoria A. Lawson (1994), 'A discussion of "Women in the Field": the Politics of Feminist Fieldwork', *Professional Geographer*, **46** (1), 96–102.

Standing, Guy (ed.) (1991), *The new Soviet labour market: in search of flexibility*, Geneva: International Labour Office.

Steyaert Chris and Rene Bouwen (1997), 'Telling Stories of Entrepreneurship – Towards a Narrative – Contextual Epistemology for Entrepreneurship Studies', in Rik Donckels and Asko Miettinen (eds), *Entrepreneurship and SME Research: On its way to the Next Millennium*, Aldershot: Ashgate, pp. 47–62.

Swain, N., A. Mihaly and Tibor Kuczi (1995), 'The Privatization of Hungarian Collective Farms', *East European Countryside*, **1**, 69–80.

Swain, Nigel (2001), 'Central European Agricultural Structures in Transition', in A. Kaleta, N. Swain, B. Weber and G. Zablocki (eds), *Revitalization of Rural Areas in Europe, Vol. IV*, Warsaw: Nicolaus Copernicus University in Toruń.

Szabó, Katalin (1992), 'Small is also beautiful in the East: the boom of small ventures in Hungary', in B.S. Katz and L. Rittenberg (eds), *The Economic Transformation of Eastern Europe. Views from Within*, Westport, CT: Praeger, pp. 127–51.

Szalai, Julia (1991), 'Some aspects of the changing situation of women in Hungary', *Signs*, **17** (1), 152–70.

Szelényi, I. (1988), *Socialist entrepreneurs: embourgeoisement in Rural Hungary*, Madison: University of Wisconsin Press.

Szelényi, I. (2001), Personal communication, Davis, CA, May.

Sziráczki, Gy. and James Windell (1992), 'Impact of employment restructuring on disadvantaged groups in Hungary and Bulgaria', *International Labour Review*, **131** (4–5), 471–96.

Szörényi, Zs. (2000), Personal communication.

Tardos, Katalin (1993), 'Férfi és Női Munkanélküliek a Somogy Megyei Településeken' ('Employment situation of men and women in rural and urban settlements of Somogy County'), *Tér és Társadalom*, **7** (1–2), 103–12.

Timár, János (1995), 'Particular features of employment and unemployment in the present stage of transformation of the Post-Socialist countries', *European Studies*, **47** (4), 633–49.

Timar, Judit (1993a), 'A Nok Tanulmányozása a Földrajzban, avagy: van-e létjogosultsága a feminista geográfiának Magyarországon?' ('Studying women in geography: or does feminist geography have grounds in Hungary?'), *Tér és Társadalom*, **7** (1–2), 1–18.

Timar, Judit (1993), 'The changing rights and conditions of women in Hungary', paper presented at the conference on 'From Dictatorship to Democracy: Women in Mediterranean, Central and Eastern Europe', 16–18 September, in M. Nash (ed.), *Proceedings*, Barcelona: University of Barcelona, pp. 1–14.

Timar, Judit (2002), 'Restructuring labour markets on the frontier of the European union: Gendered uneven development in Hungary', in A. Rainnie, A. Smith and A. Swain (eds), *Work, Employment and Transition in Post-Communist Europe*, London: Routledge, pp. 134–54.

Timar, Judit (2004), 'Gendered urban policy-making; the role of geographical scale in women's participation in Hungarian local governments', in G. Cortesi, F. Critaldi and J. Drooglever-Fortuijn (eds), *Gendered Cities. Identities, activities, networks. A life-course approach*, Rome: Societa Geografica Italiana, pp. 227–43.

Timar, Judit and G. Velkey (1998), 'Halfway between the Soviet Block and the European Union: Spatial differences in women's access to jobs in Hungary', paper presented at the Workshop on Women's rights Within the Context of European Integration, Oñati, Spain, 28–30 May.

Torres, Rebecca M. and Janet D. Momsen (2005), 'Gringolandia: the construction of a new tourist space in Mexico', *Annals of the Association of American Geographers*, **95** (2), June, 314–35.

Tóth, Andras (1992), 'The social impact of restructuring in rural areas of Hungary: Disruption of Security or the End of the Rural Socialist Middle Class Society?', *Soviet Studies*, **44** (6), 1039–43.

Tóth, Joszef (1998), 'International regional cooperation of the border areas in Hungary', in Marek Koter and Krystian Heffner (eds), *Borderlands or Transborder Regions – Geographical, Social and Political Problems, Region and Regionalism No. 3*, Opole-Lódź: Government Research Institute, Silesian Institute of Opole; Department of Political Geography and Regional Studies, University of Lódź.

Tóth, Olga (1993), 'No envy, no pity', in Funk and Mueller, *Op. cit.*, pp. 213–33.

Tykkläinen, M., E. Varis, J. Oksa, M. Piipponen, I. Nagy, E. Kiss and G. Mátray (1998), *Rural Survival Strategies in Transitional Countries. An introduction to the comparative study of localities in Northwestern Russia and Hungary*, Karelian Institute Working Paper No. 2, Joensuu, Finland: University of Joensuu.

Unicef (1999), *Women in Transition*, Regional Monitoring Report No. 6, Florence, Italy: Unicef.

UNIFEM (2002), *Progress of the World's Women 2002: Gender Equality and the United Nations Millennium Development Goals*, New York: UNIFEM.

United Nations (1992), *The Impact of Economic and Political Reform on the Status of Women in Eastern Europe: Proceedings of a United Nations Seminar*, New York: United Nations Office at Vienna, Centre for Social Development and Humanitarian Affairs.

United Nations (1993), *Women in Decision-Making: Case Study from Hungary*, New York: United Nations Office at Vienna, Centre for Social Development and Humanitarian Affairs.

United Nations (1995), *Human Development Report*, New York: United Nations.

United Nations Commission for Europe (UNECE) (2002), *Women's Entrepreneurship in Eastern Europe and CIS Countries*, Geneva: United Nations.

United States Agency for International Development (USAID) (1991), *Hungary: Gender Issues in the Transition to a Market Economy*, Washington, DC: USAID, Office of Women in Development, Bureau for Research & Development with the Bureau for Private Enterprise.

Urry, John (2005), Lecture given at the University of California, Davis, 11 May.

de Vargha, Julius (1910), *Hungary. A Sketch of the Country, its People and its Conditions*, Budapest: Printing Office of the Athenaeum.

Varis, Eira (1998), 'Facing transition in rural Hungary – a case study of an agricultural village, *Fennia*, **176** (2), 259–300.

Vasary, Ildiko (1987), *Beyond the Plan. Social Change in a Hungarian Village*, Boulder, CO: Westview Press.

Vertovec, S. (1997), 'Three meanings of "diaspora", exemplified among south Asian religions', *Diaspora*, **6**, 277–99.

Vertovec, S. (1999), 'Conceiving and researching transnationalism', *Ethnic and Racial Studies*, **22** (2), 447–77.

Voigt-Graf, C. (2004), 'Towards a geography of transnational spaces: Indian transnational commmunities in Australia', *Global Networks*, **4**, (1), 25–49.

Volgyes, Ivan and Nancy Volgyes (1977), *The Liberated Female: Life, Work and Sex in Socialist Hungary*, Boulder, CO: Westview Press.

Voszke, Eva (1995), 'Centralization, re-nationalization and re-distribution: Government's Role in changing Hungary's ownership structure', in J. Hausner, Bob Jessop and K. Nielsen (eds), *Op. cit.*, Brookfield, VT and Aldershot: Edward Elgar, pp. 287–308.

Wastl-Walter, Doris, Mónika M. Váradi and Friedrich Veider (2002), 'Bordering Silence: Border Narratives from the Austro-Hungarian Border', in Ulrike H. Meinhof (ed.), *Living (with) Borders. Identity Discourses on East-West Borders in Europe*, London: Ashgate, pp. 74–92.

Wastl-Walter, Doris and M.M. Váradi (2004), 'Ruptures in the Austro-Hungarian Border Region', in V. Pavlakovich-Kochi et al. (eds), *Op. cit.*, pp. 175–92.

Watson, Peggy (1993), 'East Europe's Silent Revolution: Gender', *Sociology*, **27** (3), 471–87.

Welter, Friederike (2002), 'Small and medium sized enterprises in Hungary', in O. Pfirrmann and G.H. Walter (eds), *Small Firms and Entrepreneurship in Central and Eastern Europe. A socio-economic perspective*, New York and Heidelberg: Physica-Verlag, pp. 139–55.

Wheelock, Jane and Elizabeth Oughton (1996), 'The Household as a Focus of Research', *Journal of Economic Issues*, **30** (1), 143–159.

Wheelock, Jane, Susan Baines, Elisabet Lunggren, Tone Magnussen, Liv Toril Pettersen and Elizabeth Oughton (1999), 'Between the household and the market: a comparative study of rural entrepreneurs in Norway and England', paper presented at the Gender and Rural Transformations: Past, Present and Future Prospects Conference, Wageningen, The Netherlands, 14–17 October.

Wong, Raymond Sin-Kwok (1995), 'Socialist stratification and mobility: Cross-national and gender differences in Czechoslovakia, Hungary and Poland', *Social Science Research*, **24** (3), 302–28.

Wong, Raymond Sin-Kwok and Robert M. Hauser (1992), 'Trends in occupational mobility in Hungary under Socialism', *Social Science Research*, **21**, (4), 419–44.

World Bank (1995), *Hungary: Structural Reforms for Sustainable Growth*, A World Bank Country Study, Washington, DC: The World Bank.

World Bank (1999), *Hungary: On the Road to the European Union*, A World Bank Country Report, Washington, DC: The World Bank.

Index

Printed and bound by CPI Group (UK) Ltd, Croydon, CR0 4YY

22/10/2024

01777640-0008